Anja Randecker

Geometry and topology of wild translation surfaces

Geometry and topology of wild translation surfaces

by
Anja Randecker

KIT Scientific Publishing

Dissertation, Karlsruher Institut für Technologie (KIT)
Fakultät für Mathematik, 2016

Tag der mündlichen Prüfung: 17. Juni 2015
Referenten: JProf. Dr. Gabriela Weitze-Schmithüsen,
Prof. Dr. Frank Herrlich, JProf. Dr. Ferrán Valdez

Impressum

KIT Scientific Publishing

Karlsruher Institut für Technologie (KIT)
KIT Scientific Publishing
Straße am Forum 2
D-76131 Karlsruhe

KIT Scientific Publishing is a registered trademark of Karlsruhe Institute of Technology. Reprint using the book cover is not allowed.

www.ksp.kit.edu

Print on Demand 2016

ISBN 978-3-7315-0456-6
DOI 10.5445/KSP/1000050964

Contents

Introduction

The subject of this thesis, wild translation surfaces, is a very recent topic in mathematics. However, the first results in the classical theory of translation surfaces go back to the early 20th century. The objects were not called translation surfaces then but had different names over the years. This is due to the fact that translation surfaces arise in very different contexts which makes them particularly interesting. There are relations to dynamical systems, Teichmüller theory, algebraic geometry, topology, geometric group theory, and many other fields.

The most visual way to define a classical translation surface is by considering finitely many polygons in the plane. If every edge of the polygons can be identified with a parallel edge of the same length so that we obtain a connected, orientable surface then the resulting object is a translation surface. It is locally flat at all points with the possible exception of the former vertices of the polygons. These exceptional points are called *singularities* and they are cone points of the resulting surface with cone angle $2\pi k$ for some $k \geq 2$.

Classical translation surfaces arise for example in the study of billiards in a polygon with rational angles, i.e. in a polygon with angles that are rational multiples of π. Fox and Kershner study billiard paths in such polygons in [FK36] by "unfolding" the polygon. This means they reflect the polygon across all edges, then reflect the images of the polygon across all images of the edges, and so on. The original, perhaps complicated billiard path in the polygon can then be considered as a straight line flow in this space of copies. To obtain a translation surface, Fox and Kershner choose a finite subset of these copies that constitute a fundamental domain under the group of reflections. The edges of the reflected polygons are identified via the product of the corresponding reflections which is in fact a translation.

The condition on the rationality of the angles is important to guarantee that the fundamental domain consists of finitely many polygons. In particular, it ensures that the singularities of the resulting translation surface are cone points as discussed above.

The unfolding construction of Fox and Kershner is described again by Katok and Zemlyakov in [ZK76] which leads to a number of articles, for instance [KMS86] by Kerckhoff, Masur, and Smillie which provides an ergodicity result on billiard paths. This is also where abelian differentials come into the theory. One obtains a translation structure on a Riemann surface by integrating over an abelian differential in neighborhoods of all points except for the zeros of the differential.

The name "translation surface", however, goes back to Thurston's lecture notes [Thu81, Chapter 3] where he defines (G, X)-manifolds and $\mathcal{G}$-manifolds without mentioning translation surfaces explicitly. In the related language of $\mathcal{F}$-structures, the translation surfaces occur again in the work of Veech. His article [Vee89] associates a group, now called Veech group, to a translation surface and relates the group to properties of the dynamics on the Teichmüller space. In [Vee90], Veech introduces the idea to use translation surfaces to estimate the volume of moduli spaces which opens up the field to algebraic geometry. Since then, the theory of classical translation surfaces has grown to a research field on its own with numerous mathematicians, including several Fields medalists, from algebraic geometry, dynamical systems, and geometric group theory working on it.

Having spoken about the classical translation surfaces (which are called *finite translation surfaces* from now on) so far, a more general concept shines through along the way. In the literature, it is often called *infinite translation surface* but as the more general object it should have the more general name – *translation surface* – in the author's view. The definition of a general translation surface is that it is a connected surface with a translation structure (see Definition 1.1).

If we do not insist, for example, on rational angles then the unfolding of a polygon provides an infinite translation surface, studied by Valdez in [Val09] and [Val12]. As we have indicated in the discussion before, the singularities of the resulting translation surface do not need to be cone points.

Another possibility to produce examples is to consider infinite coverings of finite translation surfaces as was done by Hubert and Schmithüsen in [HS10], by Hooper, Hubert, and Weiss in [HW12], [HHW13], and [HW13], and by Frączek and Ulcigrai in [FU14], for example. A special example of this class of translation surfaces is the wind-tree model which goes back to a model of Ehrenfest and Ehrenfest that describes the movement of gaseous particles. Many different aspects of the wind-tree model are studied, for instance, by Avila, Delecroix, Hubert, Lelièvre, and Troubetzkoy in [DHL14], [HLT11], and [AHar].

In the two mentioned constructions, the singularities of the translation surfaces are not cone points in the strict sense any more but they can be considered as infinite angle analogs of cone points. For both kinds of cone points it is true that there exists a punctured neighborhood of the singularity which is a finite or infinite cyclic translation covering of the once-punctured flat disk.

In [Cha04], Chamanara describes a first example of a family of translation surfaces so that each translation surface in that family has a singularity which has no punctured neighborhood that is a translation covering of a once-punctured disk. In particular, every neighborhood of a singularity in that family has infinite genus and so the topology of the corresponding translation surface is more complicated but also raises more interesting questions. Singularities of this type are called *wild* and translation surfaces with wild singularities are called *wild translation surfaces.*

Different examples of wild translation surfaces have then been studied, for instance arising as a geometric limit of the Arnoux-Yoccoz surfaces defined by Bowman in [Bow13] and arising from a generalization of Thurston's construction (see [Thu88]) by Hooper in [Hoo15]. However, the systematic study of wild singularities started quite recently with the work of Bowman and Valdez. In [BV13], they define *linear approaches* and *rotational components* (see Section 1.2 for definitions) which makes it possible to study a wild singularity in terms of geodesic rays emanating from the singularity.

In the theory of finite translation surfaces, an easy-to-prove but fascinating equation is true: the *Gauß-Bonnet formula* provides a correlation between the topology of translation surfaces and the geometry of their singularities. In fact, the equation relates the genus and the cone angle of the singularities, without using any other information (see Proposition 1.14). A similar result on general translation surfaces would not only be nice in itself but could help to classify translation surfaces. For finite translation surfaces, there exists a well-established theory on moduli spaces and strata of translation surfaces which has recently made a big leap with the results of Eskin, Mirzakhani, and Mohammadi in [EMM15] and of Eskin and Mirzakhani in [EM13]. Such a theory is desirable for general translation surfaces but there is still a lack of a good notion of moduli spaces and even more of strata of translation surfaces. To remedy this lack it seems natural to first study singularities (especially wild singularities) in more detail and establish relations between the singularities as a geometric object of a translation surface and the topology in terms of genus and ends.

The purpose of this thesis is to contribute to these new aspects of the theory. We do this under three different points of view. As one aspect, we study the local neighborhood of a singularity, in terms of topological properties and their spaces of rotational components. As a second aspect we consider the global topology of wild translation surfaces, in particular the spaces of ends that can occur. The third aspect is to bring together these two points of view and show under which conditions the existence of a singularity with given properties can imply topological properties of the translation surface.

The structure of this thesis is as follows. In Chapter 1, we recall the basic definitions and concepts that are needed in the course of the thesis. We start with a discussion of definitions of (finite) translation surfaces and types of singularities. Then the spaces of linear approaches and rotational components are defined. The classification of closed and of general surfaces is recapped, together with the definition of the space of ends that is needed for the classification. The chapter concludes with basic properties of the Veech group of a translation surface and of the geodesic flow on a translation surface.

The recalled concepts are enhanced in Chapter 2 by studying three examples in detail, i.e. by determining their geometry in terms of Veech groups, types of singularities, spaces of linear approaches, and spaces of rotational components, and by determining their topology in terms of genus and space of ends. Moreover, we give a general construction of modifying translation surfaces so that they are provided with additional singularities and desirable properties.

In Chapter 3, the theory of ends of a surface is specialized to a theory of ends of translation surfaces. This includes the definition of singular and regular ends and the end of a singularity. In Theorem 1, we prove that almost every combination of genus, ends, and singularities can be realized by a translation surface.

Theorem 1 (No restrictions in terms of topology and number of singularities)
Let (s, g, e) be a triple of natural numbers with $s \geq 1$, $g \geq 0$, and $e \geq 1$. Then there exists a translation surface with s (nonremovable) cone angle or infinite angle singularities, genus g, and e ends.

Moreover, in Theorem 2, which was proven together with Camilo Ramírez Maluendas, we show that all spaces of ends of surfaces also occur as spaces of ends of translation surfaces.

Theorem 2 (Existence of a translation surface with given ends)
Let E, E' be closed subsets of the Cantor set with $E' \subseteq E \neq \emptyset$. Then there exists a translation surface $(X, \mathcal{A})$ that has only cone angle singularities, i.e. the complex structure of X can be extended to $\overline{X}$, and such that $\mathrm{Ends}(\overline{X})$ is homeomorphic to E and $\mathrm{Ends}_{nonplanar}(\overline{X})$ is homeomorphic to E'.

We then establish relations between the ends and singularities, summarized in Theorem 3.

Theorem 3 (Relations between ends and singularities)
Let $(X, \mathcal{A})$ be a translation surface, e an end of $(X, \mathcal{A})$ and σ a singularity of $(X, \mathcal{A})$.

(i) If e is a regular end then it is planar and of finite area.

(ii) If e is a singular end then it can be induced by every number of singularities, including zero.

(iii) If σ is a cone angle singularity then $\mathrm{end}(\sigma)$ is planar and of finite area.

(iv) If σ is an infinite angle singularity then $\mathrm{end}(\sigma)$ is of infinite area and can be planar or nonplanar.

(v) If $(X, \mathcal{A})$ has finitely many singularities and e is a singular, planar end of finite area then e is induced by a singularity.

(vi) If $\overline{X}$ is compact then the ends and the singularities are in one-to-one-correspondence.

In Chapter 4, we study the neighborhoods of singularities, in particular of wild singularities. This is done in terms of geometric and topological properties and in terms of saddle connections. In particular, we define the property *xossiness* for singularities which indicates if there exist saddle connections of arbitrarily small length that have a nice neighborhood. Two conditions are given under which xossiness is fulfilled.

We use the results on xossiness to show in Chapter 5 under which conditions the existence of wild singularities has influence on the genus. In particular, we give a characterization of infinite genus in terms of saddle connections and exhibit in Theorem 4 that the existence of a wild singularity which fulfills xossiness together with the existence of two recurrent directions for the geodesic flow implies infinite genus.

Theorem 4 (Wild singularity implies infinite genus)
Let $(X, \mathcal{A})$ be a translation surface so that for two directions $\theta_1, \theta_2 \in S^1$ the geodesic flows F_{θ_1} and F_{θ_2} are recurrent. Furthermore, let σ be a wild singularity of $(X, \mathcal{A})$ that fulfills xossiness. Then X has infinite genus.

The results in this chapter, together with some statements in the previous chapter have been published before in [Ran14].

The results in Chapter 6 are all about the spaces of rotational components and about their connections to the spaces of linear approaches and the translation structure on rotational components. Three relations on the different topologies are established. We show in Theorem 5 that the space of rotational components does not at all determine the space of linear approaches.

Theorem 5 ($\widetilde{\mathcal{L}}(\sigma)$ does not determine $\mathcal{L}(\sigma)$)
There are uncountably many translation structures $\mathcal{A}_r$ on a Loch Ness monster of finite area with the following properties.

(i) Every translation surface $(X_r, \mathcal{A}_r)$ has exactly one singularity σ_r.

(ii) The spaces $\widetilde{\mathcal{L}}(\sigma_r)$ of rotational components are all homeomorphic.

(iii) The spaces $\mathcal{L}(\sigma_r)$ of linear approaches are pairwise not homeomorphic.

In the last theorem, Theorem 6, we show that every finite topological space can be realized as the space of rotational components of some wild singularity.

Theorem 6 (Every finite space occurs as $\widetilde{\mathcal{L}}(\sigma)$)
Let Y be a non-empty topological space of finite cardinality. Then there exists a translation surface $(X, \mathcal{A})$ with a singularity σ so that $\widetilde{\mathcal{L}}(\sigma)$ carries the same topology as Y.

These results as well as some findings on examples from Chapter 2 were obtained together with Lucien Clavier and Chenxi Wu and have been published before in [CRW14].

At this point, I would like to express my gratitude to the various people and institutions that supported my work. First of all, I want to heartily thank my advisor *Gabi Weitze-Schmithüsen* for her constant encouragement and the tremendous amount of time she took to discuss translation surfaces and the life in academia. Also, I am much obliged to *Frank Herrlich* and *Ferrán Valdez* for agreeing to become referees of this thesis. *Frank Herrlich* I also want to thank for his famous "open door" and the frequent casual discussions in the evening whereas I want to thank *Ferrán Valdez* for encouraging me a lot by being interested in my progress and sharing his thoughts on wild translation surfaces every time we met. I also want to thank all previously mentioned, *Stefan Kühnlein*, *Pat Hooper*, *Chenxi Wu*, *Lucien Clavier*, and all members of the *Kaffeerunde* who taught me different fields of mathematics, in lectures, in discussions, by answering my questions, or by working on a joint article. In particular, I want to thank my office mates *Myriam Finster*, *Florian Nisbach*, *Benni Peters*, and *Sven Caspart* for the lively interchange of concepts in mathematics and culinary art.

I am also indebted to the *Institute for Computational and Experimental Research in Mathematics (ICERM)* that funded a semester-long research stay in Providence during the program "Low-dimensional Topology, Geometry, and Dynamics" and to the *Karlsruhe House of Young Scientists (KHYS)* that supported me with two grants which made it possible to visit Pat Hooper in New York and to invite Chenxi Wu and Lucien Clavier to Karlsruhe. Finally, I want to thank *Gabi Weitze-Schmithüsen*, *Ferrán Valdez*, *Frank Herrlich* and *Myriam Finster* for proofreading and giving valuable advice on how to improve the exposition.

1 Basics on translation surfaces

In this chapter, we recall the definitions of translation surfaces, of their singularities and their space of ends, of Veech groups and of dynamics on translation surfaces. The fundamental properties are stated, mostly without proof because detailed proofs are covered in various sources. Regarding the classical theory of translation surfaces, there are a lot of elaborate surveys with different emphases, for instance [MT02], [Zor06], and [FM13]. Most of the following has also been previously formulated in [Ran12].

1.1 Translation surfaces

We start with the definition and some examples of our main object.

Definition 1.1 (Translation surface)
A *translation surface* $(X, \mathcal{A})$ is a connected two-dimensional manifold X together with a translation structure $\mathcal{A}$ on X, i.e. a maximal atlas on X so that the transition functions are locally translations.

In particular, translation surfaces are Riemann surfaces, hence triangulable and second-countable.

A trivial example of a translation surface is the Euclidean plane $\mathbb{R}^2$. The charts of the translation atlas are given by the identity and by maps on open sets in $\mathbb{R}^2$ that are locally translations. Going one step towards non-triviality, we see that the quotients $\mathbb{R}^2/\mathbb{Z}$ (which is a cylinder) and $\mathbb{R}^2/\mathbb{Z}^2$ (which is a torus) are also translation surfaces. We can show this explicitly by specifying charts for the cylinder or the torus. To define a chart, consider a simply connected subset of the cylinder or the torus and choose a connected component of the preimage in $\mathbb{R}^2$ under the quotient map π. Then the restriction of π^{-1} to the chosen set is a chart.

Let $(X, \mathcal{A})$ be a translation surface and $p\colon U \to \mathbb{R}^2$ a chart such that U is simply connected. Then U is homeomorphic to a connected subset of $\mathbb{R}^2$ and we can pull back the Euclidean metric from $\mathbb{R}^2$ to U. We define the metric on X as the path length metric induced by the local metrics on the charts of this type.

Every metric space has a *metric completion* which is isometric to the space of all Cauchy sequences modulo an equivalence relation. Here, two Cauchy sequences are considered to be equivalent if and only if the sequence of the distances between the corresponding elements converges to 0. As we can see from this construction, the metric completion is a complete metric space with respect to the extended metric. For the Euclidean plane, the cylinder, and the torus, the metric completion is equal to the original space. For a punctured metric space, the metric completion has additional points which correspond to filling in the punctures. This phenomenon in particular occurs for the class of translation surfaces which is defined in the next definition.

Definition 1.2 (Finite translation surface)
A translation surface $(X, \mathcal{A})$ is called *finite* if the metric completion $\overline{X}$ is a compact surface and $\overline{X} \setminus X$ is discrete.

As X is a surface and hence is open in $\overline{X}$ the set $\overline{X} \setminus X$ is closed. Therefore it is discrete if and only if it is finite. The additional points in the metric completion will be the starting point for our studies.

Definition 1.3 (Singularity)
For a translation surface $(X, \mathcal{A})$, every element of $\overline{X} \setminus X$ is called *singularity* of $(X, \mathcal{A})$.

The Euclidean plane and the cylinder are not finite translation surfaces whereas the torus is. However, all three of them do not have singularities.

Definition 1.2 is not among the three standard definitions for a finite translation surface. A frequently used standard definition requires a description of the singularities in terms of translation coverings.

Definition 1.4 (Translation covering)
A continuous map $p\colon (X, \mathcal{A}) \to (Y, \mathcal{B})$ of translation surfaces is called *translation covering* if it is a covering map such that the translation structure $\mathcal{A}$ on X is the pullback of the translation structure $\mathcal{B}$ on Y.

In the case of a translation covering $p\colon (X, \mathcal{A}) \to (Y, \mathcal{B})$, the deck transformations of p act as translations on $(X, \mathcal{A})$, i.e. for every deck transformation f and every $x \in X$ there exist charts $(U, \varphi), (V, \psi) \in \mathcal{A}$ with $x \in U$

and $f(U) \subseteq V$ so that for every $z \in \varphi(U) \subseteq \mathbb{R}^2$ it is true that $\psi \circ f \circ \varphi^{-1}$ is a translation.

Definition 1.5 (Finite translation surface)
Let $\widehat{X}$ be a closed, connected Riemann surface with a finite set $\Sigma \subseteq \widehat{X}$ of marked points and a translation structure $\mathcal{A}$ on $X := \widehat{X} \setminus \Sigma$. Furthermore, for every $\sigma \in \Sigma$, there shall exist an $\epsilon > 0$ and a continuous map $p\colon B(\sigma, \epsilon) \setminus \{\sigma\} \to B(0, \epsilon) \setminus \{0\} \subseteq \mathbb{R}^2$ such that p is a finite cyclic translation covering of the once-punctured disk. Then $(X, \mathcal{A})$ is called a *finite translation surface.*

We have to show that the two given definitions of finite translation surfaces coincide.

Proposition 1.6 (Finite translation surfaces are finite translation surfaces)
Every finite translation surface according to Definition 1.2 is a finite translation surface according to Definition 1.5 and vice versa.

Proof. Let $(X, \mathcal{A})$ be a finite translation surface according to Definition 1.2. We define $\widehat{X} := \overline{X}$ and $\Sigma := \overline{X} \setminus X$. Then $\widehat{X}$ is a connected, closed topological surface and Σ is finite. Moreover, X carries a translation structure, hence a complex structure which can be extended to a complex structure on $\widehat{X}$ by Picard's great theorem. Furthermore, let $\sigma \in \Sigma$. Then there exists an $\epsilon > 0$ such that $B(\sigma, \epsilon) \subseteq \widehat{X}$ is homeomorphic to a disk, in particular to $B(0, \epsilon) \subseteq \mathbb{R}^2$. Since $B(\sigma, \epsilon) \setminus \{\sigma\}$ is locally flat, we can define a map $p\colon B(\sigma, \epsilon) \setminus \{\sigma\} \to B(0, \epsilon) \setminus \{0\}$ by composing charts (concatenated with transition functions where necessary). This is locally an isometry and moreover, it is a cyclic translation covering. Hence, $(X, \mathcal{A})$ is a finite translation surface according to Definition 1.5.

Now, let $(X, \mathcal{A})$ be a finite translation surface according to Definition 1.5. The metric completion $\overline{X}$ is equal to the compact surface $\widehat{X}$ and $\overline{X} \setminus X = \Sigma$ is a discrete set. Hence, $(X, \mathcal{A})$ is a finite translation surface according to Definition 1.2. □

There are two more equivalent definitions for finite translation surfaces (cf. [Mas06, Definition 1, 4, and 5]). The first one defines finite translation surfaces as Riemann surfaces with a nonzero abelian differential on it. The translation structure is obtained by integrating over the differential in neighborhoods of all points except for the zeros. The zeros of the differential yield the singularities of the translation surface.

The second one is given by a construction which can be extended to the more general situation: Consider a set of finitely many disjoint polygons in the Euclidean plane so that for every edge of a polygon there exists a parallel edge of the same length. Now we identify the matching edges of the polygons via translations. If the result is a connected, orientable surface then it is a finite translation surface. We have a translation chart on the neighborhood of every point except possibly for the vertices. Therefore, we consider the resulting surface without the former vertices as a translation surface. In that way, the former vertices are exactly the additional points in the metric completion and so they are the singularities of the translation surface.

The easiest example is that of a square where top and bottom are identified via a translation and left edge and right edge are identified via a translation which yields a torus with translation structure.

We consider three more examples that are obtained by gluing polygons to get used to that construction. We will consult these examples for the clarification of definitions throughout the following sections.

The first example is very classical and explicitly appears for the first time in [Vee89, Section 4].

Example 1.7 (Veech's double n-gons). Let $n \geq 3$ and consider two regular n-gons of side length 1 without vertices for which two edges are parallel. The two edges have to be chosen so that there is one of the edges in each n-gon and so that the polygons are on different sides of the edges. Then for every edge there also exists a parallel edge in the other n-gon. These parallel edges are now identified via the corresponding translations. In Figure 1.1 we have $n = 5$ and by labeling two edges with the same letter we indicate that they are identified.

As the resulting surface is orientable, connected and obtained by gluing finitely many polygons, it is a finite translation surface. We can also check by hand that the metric completion is a compact surface of genus $\frac{n-1}{2}$ if n is odd and a compact surface of genus $\frac{n-2}{2}$ if n is even. In the case of n odd, the number of singularities is 1 whereas in the case of n even it is 2, hence it is finite in both cases.

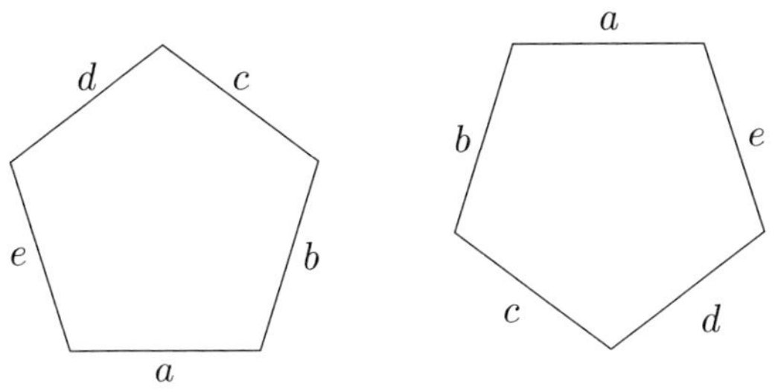

Figure 1.1: Veech's double n-gon for $n = 5$: edges which are labeled by the same letter are identified.

We can not only glue finite translation surfaces from finitely many polygons but also obtain a translation surface when gluing infinitely many polygons in the same way as described above. This is used in the next two examples.

The following gives an example of a translation surface which is not finite but still not too complicated. It is studied in [HS10] as an exceptional member of a family of translation surfaces and in [HHW13] because of its interesting dynamics.

Example 1.8 (Infinite staircase). Consider infinitely many copies of a square without vertices. The squares are arranged as shown in Figure 1.2 and again the gluings are indicated by letters.

The metric completion of this surface has infinite area and is not compact, so the infinite staircase is not a finite translation surface.

The last example for now is described in detail in [Cha04] which is one of the first articles dealing with a translation surface which is not a finite translation surface and not a translation covering of a finite translation surface. Sometimes the surface is called *baker's map surface* as it can be obtained from a dynamical system given by a chaotic map from the unit square into itself which is called *baker's map*.

Example 1.9 (Chamanara surface). We consider a square without vertices and want to specify gluings to make the square into a translation surface. For this, top and bottom have to be identified. Doing the identification in the canonical way we obtain a torus as described before. However, we do the identification in the following non-canonical way. We divide the top

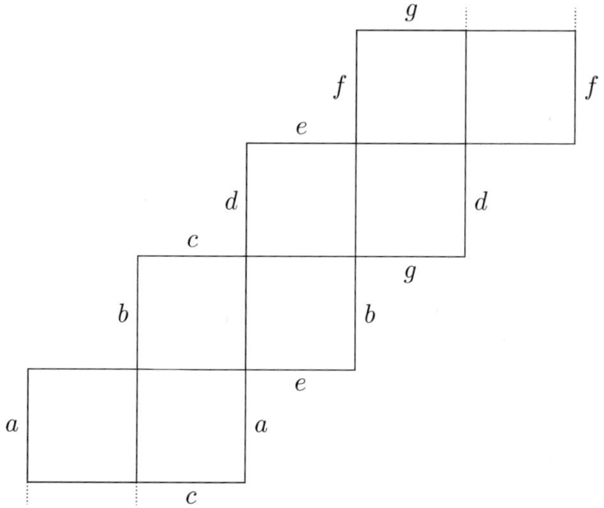

Figure 1.2: For the infinite staircase, opposite edges are identified.

edge into two halves and the bottom edge into two halves. Then we glue the right half of the top to the left half of the bottom. The remaining halves are divided again and the right part of the top is glued to the left part of the bottom, and so on (see Figure 1.3).

We do the same with the left and the right edge, always identifying the upper part of the right edge with the lower part of the left edge.

When we exclude not only the vertices of the square but also the points on the edges where we divided the segments into halves, we obtain a translation structure on the resulting surface. The vertices and the cutting points lead to points in the metric completion and so they are forming singularities. We will discuss the singularities of the Chamanara surface in Example 1.15 and in Section 2.1 in more detail.

To complete this section, we introduce two more definitions that are useful to describe translation surfaces.

Definition 1.10 (Saddle connection)
Let $(X, \mathcal{A})$ be a translation surface. A *saddle connection* of $(X, \mathcal{A})$ is a

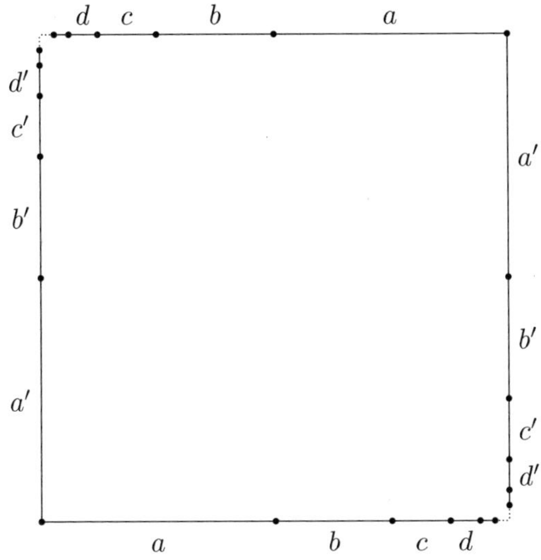

Figure 1.3: For the Chamanara surface, we identify segments that are parallel and have the same length.

geodesic in $\overline{X}$ which joins two (not necessarily different) singularities and does not contain a singularity in its interior.

A saddle connection defines a vector in $\mathbb{R}^2$ by its length and direction. More generally, for every geodesic we can define a holonomy vector, using the *developing map* from the universal cover of the translation surface to $\mathbb{R}^2$ (see [Rat94, § 8.4] for a definition of the developing map).

Definition 1.11 (Holonomy vector)
Let $(X, \mathcal{A})$ be a translation surface and $\gamma\colon [0,l] \to \overline{X}$ a geodesic with $\gamma((0,l)) \subseteq X$. Using the charts of the translation atlas, we can develop $\gamma((0,l))$ into $\mathbb{R}^2$ and obtain a geodesic in $\mathbb{R}^2$. The difference vector between the start and the end point of the geodesic in $\mathbb{R}^2$ is called *holonomy vector* of the geodesic γ.

It is shown in [Vor96, Proposition 3.1] that the set of holonomy vectors of saddle connections of a finite translation surface is discrete in $\mathbb{R}^2$.

1.2 Singularities of translation surfaces

We defined the singularities of a translation surface $(X, \mathcal{A})$ in Definition 1.3 as the additional points in the metric completion $\overline{X}$. These additional points provide rich information on the geometry and topology of the surface if there are not too many of them. Considering the open disk $B(0,1) \subseteq \mathbb{R}^2$ as a translation surface, we obtain that the set of singularities is isometric to $S^1 = \partial B(0,1)$ with the metric inherited from $\mathbb{R}^2$. Most authors want to exclude this behaviour and we do the same by restricting ourselves to translation surfaces with discrete sets of singularities.

Convention (Translation surfaces have discrete sets of singularities). From now on, by a translation surface we always mean a translation surface where the set of singularities is discrete in the metric completion.

In this section we distinguish different types of singularities and present an approach how to describe singularities, following [BV13].

As we have seen in Proposition 1.6, the singularities of finite translation surfaces have a nice description in terms of translation coverings of a once-punctured disk. We distinguish singularities that have such a description from singularities that have not.

Definition 1.12 (Cone angle, infinite angle, and wild singularity) Let $(X, \mathcal{A})$ be a translation surface and σ a singularity of $(X, \mathcal{A})$.

(i) The singularity σ is called *cone angle singularity of multiplicity* $k \geq 1$ if there exist

- $\epsilon > 0$,
- an open neighborhood B of σ in $\overline{X}$, and
- a k-cyclic translation covering from $B \setminus \{\sigma\}$ to the once-punctured disk $B(0,\epsilon) \setminus \{0\} \subseteq \mathbb{R}^2$.

If $k = 1$ then the singularity σ is also called *removable singularity* or *flat point.*

(ii) The singularity σ is called *infinite angle singularity* or *cone angle singularity of multiplicity* ∞ if there exist

- $\epsilon > 0$,
- an open neighborhood B of σ in $\overline{X}$, and

- an infinite cyclic translation covering from $B \setminus \{\sigma\}$ to the once-punctured disk $B(0,\epsilon) \setminus \{0\} \subseteq \mathbb{R}^2$.

(iii) The singularity σ is called *wild* if it is neither a cone angle nor an infinite angle singularity.

Definition 1.13 (Wild and tame translation surface)
A translation surface is called *wild* if it has a wild singularity. If it has no singularities or if all singularities are cone angle or infinite angle singularities then the translation surface is called *tame*.

We have seen in Proposition 1.6 that every finite translation surface is a tame translation surface. For instance, the unique singularity of the double pentagon in Example 1.7 has a punctured neighborhood which is a triple covering of the once-punctured disk (cf. Figure 1.4). Hence, it is a cone angle singularity of multiplicity 3.

Also, the infinite staircase in Example 1.8 is a tame translation surface since it has four singularities as indicated in Figure 1.5 of which all of them are infinite angle singularities.

For a finite translation surface, the *Gauß-Bonnet formula* states that all topological information is encoded in the number and multiplicities of the singularities. The proof is very short and fits nicely in the discussion of translation surfaces obtained by gluing polygons, so we include it here.

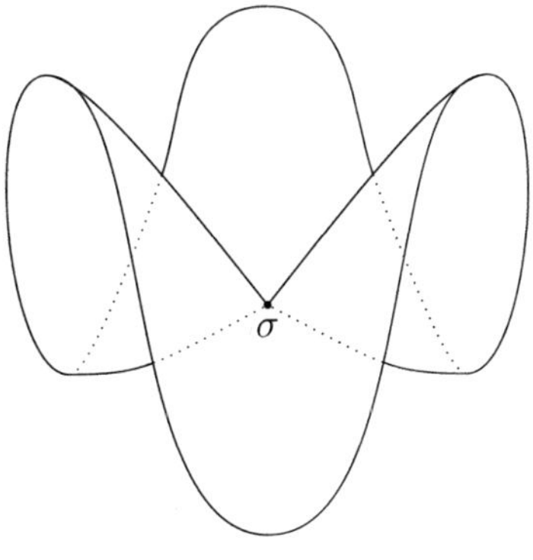

Figure 1.4: A metric picture of the neighborhood of a cone angle singularity σ of multiplicity 2.

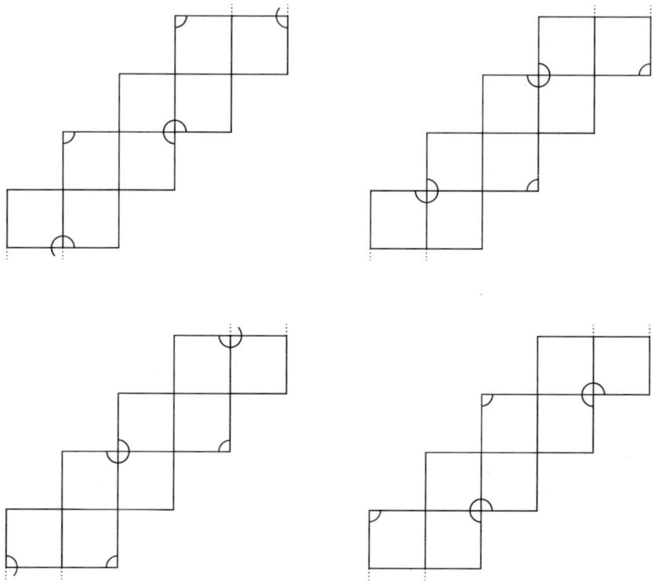

Figure 1.5: In each copy of the infinite staircase, one of the four infinite angle singularities is indicated.

Proposition 1.14 (Gauß-Bonnet formula)
Let $(X, \mathcal{A})$ be a finite translation surface of genus g that has n cone angle singularities of multiplicity k_i, $i = 1, \ldots, n$. Then it holds

$$g = 1 + \frac{1}{2}\sum_{i=1}^{n}(k_i - 1).$$

Proof. Consider a triangulation of $\overline{X}$ so that all singularities are vertices of triangles. Let v be the number of vertices, e the number of edges and f the number of faces. When we sum up the angles of all triangles in two different ways, we obtain

$$f \cdot \pi = \sum_{i=1}^{n} k_i \cdot 2\pi + (v - n) \cdot 2\pi.$$

Together with $2e = 3f$, this leads to

$$2 - 2g = \chi\left(\overline{X}\right) = v - e + f = v - \frac{1}{2}f = -\sum_{i=1}^{n}(k_i - 1). \qquad \square$$

The Gauß-Bonnet formula shows in particular that the torus is the only surface that can be equipped with a translation structure so that we obtain a finite translation surface without singularities.

Informally, the multiplicity of a cone angle singularity can be described by counting "how many times we surround the singularity" in a given distance in the flat metric before the curve closes. We would like to have a similar way of describing a wild singularity. An example shall point out the problems we have to deal with.

Example 1.15 (Singularities of the Chamanara surface). Recall the Chamanara surface from Example 1.9 where the edges of a square are divided into segments and these segments are glued in a crosswise way. By surrounding the cutting points via the gluings we find that every second cutting point is identified as sketched in Figure 1.6. However, the distance of the two indicated points in the figure is not bounded away from 0 in the metric completion, hence it is 0. This means that all cutting points

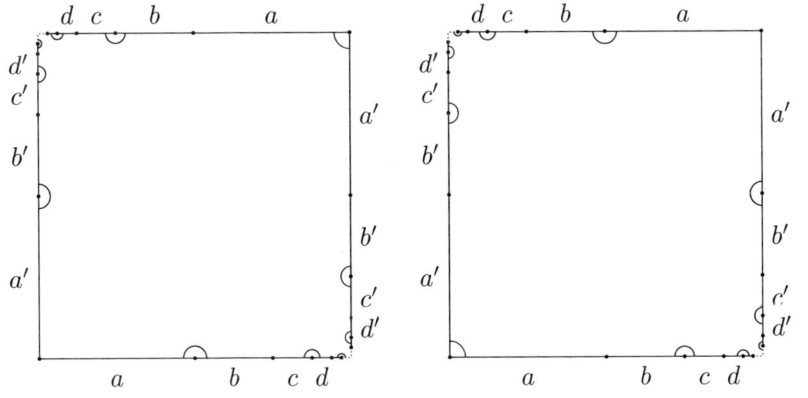

Figure 1.6: The arcs indicate which cutting points are identified by the gluings in the Chamanara surface.

and the lower left and upper right corner are identified to one point in the metric completion. The same argument holds true for the upper left and lower right corner of the square and we receive that we have exactly one singularity σ.

We observe in Figure 1.6 that for every $\epsilon > 0$, the ϵ-neighborhood of σ contains a saddle connection and is not simply connected. Hence $B(\sigma, \epsilon) \setminus \{\sigma\}$ is not a translation covering of a once-punctured disk. In an informal way, we can say that the singularity itself is an obstacle to the existence of a translation covering. So, σ is neither a cone angle nor an infinite angle singularity, which means it is a wild singularity.

We can see in the last example that, informally spoken, there are indeed possibilities to go around the singularity, for example by following the arcs in Figure 1.6. However, when doing this we will never see the whole complexity of the singularity: we have two disjoint possibilities to go around the singularity and still will not reach the upper left and lower right corner.

So, to understand wild singularities better we have to describe them more detailed. The first approach of that was done by Bowman and Valdez in [BV13] by introducing *linear approaches* and *rotational components.* We recall their definitions and some useful properties in the remainder of this section.

The first definition deals with all possibilities of leaving a singularity in a geodesic way. Recall that a *geodesic curve* is a continuous map $\gamma\colon I \to X$ ($I \subseteq \mathbb{R}$ an interval) so that for each $t \in I$ there exists a neighborhood $(t_1, t_2) \subseteq I$ of t so that $\gamma_{|(t_1,t_2)}$ is isometric. In particular, this means that all geodesic curves are unit speed.

Definition 1.16 (Space of linear approaches)
Let $(X, \mathcal{A})$ be a translation surface, $x \in \overline{X}$, and $\epsilon > 0$. We define

$$\mathcal{L}^\epsilon(X) := \{\gamma\colon (0,\epsilon) \to X : \gamma \text{ is a geodesic curve}\}$$

and

$$\mathcal{L}^\epsilon(x) := \left\{\gamma \in \mathcal{L}^\epsilon(X) : \lim_{t\to 0} \gamma(t) = x\right\}.$$

If x is a wild singularity then we can deduce from Definition 1.12 that there exists no $\epsilon > 0$ so that all geodesic curves starting in x can be extended to have at least length ϵ. Therefore, we consider equivalence classes instead of curves: $\gamma_1 \in \mathcal{L}^\epsilon(X)$ and $\gamma_2 \in \mathcal{L}^{\epsilon'}(X)$ are called *equivalent* if $\gamma_1(t) = \gamma_2(t)$ for every $t \in (0, \min\{\epsilon, \epsilon'\})$. The space

$$\mathcal{L}(X) := \bigsqcup_{\epsilon>0} \mathcal{L}^\epsilon(X)/\sim$$

is called *space of linear approaches of* X.

In the same way, we define

$$\mathcal{L}(x) := \bigsqcup_{\epsilon>0} \mathcal{L}^\epsilon(x)/\sim\ \subseteq \mathcal{L}(X)$$

as the *space of linear approaches of* x and the equivalence class $[\gamma]$ of $\gamma \in \mathcal{L}^\epsilon(x)$ is called *linear approach* to the point x.

Every $\mathcal{L}^\epsilon(X)$ can be embedded in $\mathcal{L}(X)$ and also in $\mathcal{L}^{\epsilon'}(X)$ for every $\epsilon' > 0$ with $\epsilon \geq \epsilon'$. The family of all spaces $\mathcal{L}^\epsilon(X)$ together with these embeddings is a *direct system*, i.e. the composition of two such embeddings $\mathcal{L}^\epsilon(X) \hookrightarrow \mathcal{L}^{\epsilon'}(X)$ and $\mathcal{L}^{\epsilon'}(X) \hookrightarrow \mathcal{L}^{\epsilon''}(X)$ is equal to the given embedding $\mathcal{L}^\epsilon(X) \hookrightarrow \mathcal{L}^{\epsilon''}(X)$. We have that $\mathcal{L}(X)$ is the colimit of the described direct system.

For a regular point $x \in X$, the space $\mathcal{L}(x)$ of linear approaches of x is in one-to-one correspondence to a space $\mathcal{L}^\epsilon(x)$ for some $\epsilon > 0$. Therefore, the space $\mathcal{L}(x)$ of linear approaches corresponds to the set of rays in all directions in $S^1 = \mathbb{R}/2\pi\mathbb{Z}$ so it is in one-to-one correspondence to the fibre of the unit tangent bundle at x. For now, we are considering the spaces of linear approaches only as sets. To compare the spaces of linear approaches of different points, we want to establish a topology on $\mathcal{L}(X)$.

Before describing the topology on $\mathcal{L}(X)$, we define rotational components as classes of linear approaches with a one-dimensional translation structure on it. This is done by mimicking the concept of "going around a singularity" for cone angle and infinite angle singularities.

For $\epsilon > 0$ and a generalized interval I, i.e. a nonempty connected subset of $\mathbb{R}$, we consider the infinite strip $\{z \in \mathbb{C} : \mathrm{Re}(z) < \log\epsilon, \mathrm{Im}(z) \in I\} \subseteq \mathbb{C}$. Via the injective map

$$f\colon \mathbb{C} \to (\mathbb{C} \setminus \{0\}) \times \mathbb{R},\ z \mapsto (e^z, \mathrm{Im}(z))$$

we can spiral the strip around the puncture at 0. We endow the image U of the strip under f with the pullback of the Euclidean metric on $\mathbb{C}$ via

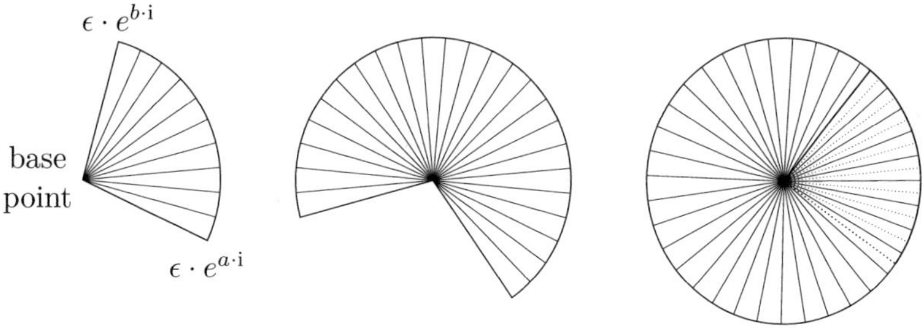

Figure 1.7: Examples for $i_\epsilon(U)$ with $I = [a, b]$: in the first two examples we have $b - a < 2\pi$ whereas in the last example we have $b - a > 2\pi$.

the projection to the first component. Some examples of the projection of U are sketched in Figure 1.7.

Such an image U together with an embedding in X defines an angular sector as is made precise in the following definition.

Definition 1.17 (Angular sector)
Let $(X, \mathcal{A})$ be a translation surface. An *angular sector* is a triple $(I, \epsilon, i_\epsilon)$ of a generalized interval I, $\epsilon > 0$, and an isometric embedding i_ϵ of

$$U := f\left(\{z \in \mathbb{C} : \operatorname{Re}(z) < \log \epsilon, \operatorname{Im}(z) \in I\}\right)$$

into X.

For an angular sector $(I, \epsilon, i_\epsilon)$ and $y \in I$, $\lim_{x \to -\infty}(i_\epsilon \circ f)(x + \mathrm{i}y)$ is a point in $\overline{X}$ and independent of y. This point is called *base point* of the angular sector $(I, \epsilon, i_\epsilon)$.

If $(I, \epsilon, i_\epsilon)$ is an angular sector with base point $x \in \overline{X}$ then we can define a map $f_{(I,\epsilon,i_\epsilon)}\colon I \to \mathcal{L}(x)$. For every $y \in I$, the image of the map

$$i_\epsilon \circ f\colon \{z \in \mathbb{C} : \operatorname{Re}(z) < \log \epsilon, \operatorname{Im}(z) = y\} \to X$$

is a geodesic segment of length ϵ and hence induces an element in $\mathcal{L}^\epsilon(x)$. Let $f_{(I,\epsilon,i_\epsilon)}(y)$ be the corresponding linear approach in $\mathcal{L}(x)$.

In the next definition, we use angular sectors to define an equivalence relation on the space $\mathcal{L}(x)$ of linear approaches of a point x. In an informal way, we can describe it as two linear approaches being "contained" in the image of a given angular sector.

Definition 1.18 (Rotational component)
Let $(X, \mathcal{A})$ be a translation surface, $x \in \overline{X}$, and $[\gamma_1], [\gamma_2] \in \mathcal{L}(x)$. The linear approaches $[\gamma_1]$ and $[\gamma_2]$ are called *R-equivalent* if there exists an angular sector $(I, \epsilon, i_\epsilon)$ with base point x and $y_1, y_2 \in I$ such that $f_{(I,\epsilon,i_\epsilon)}(y_1) = [\gamma_1]$ and $f_{(I,\epsilon,i_\epsilon)}(y_2) = [\gamma_2]$.

The R-equivalence class $\overline{[\gamma]}$ of $[\gamma] \in \mathcal{L}(x)$ is called *rotational component* of x.

We say that a linear approach $[\gamma]$ is *contained* in a rotational component c if $\overline{[\gamma]} = c$ holds.

The set of all linear approaches assigned to a given angular sector $(I, \epsilon, i_\epsilon)$ can be described as

$$V(I, \epsilon, i_\epsilon) := \left\{ f_{(I,\epsilon,i_\epsilon)}(y) : y \in I \right\}.$$

Then $V(I, \epsilon, i_\epsilon)$ is a subset of the space $\mathcal{L}(x)$ of linear approaches. All linear approaches in $V(I, \epsilon, i_\epsilon)$ are contained in the same rotational component c where $c := \overline{f_{(I,\epsilon,i_\epsilon)}(y)}$ for an arbitrary $y \in I$. In other words, the map

$$\varphi_{(I,\epsilon,i_\epsilon)} \colon V(I, \epsilon, i_\epsilon) \to \mathbb{R}, \ f_{(I,\epsilon,i_\epsilon)}(y) \mapsto y$$

is inverse to $f_{(I,\epsilon,i_\epsilon)}$ and hence $V(I, \epsilon, i_\epsilon)$ and I are in one-to-one correspondence.

We want to define a topology on a rotational component $\overline{[\gamma]}$ of x by considering all angular sectors $(I, \epsilon, i_\epsilon)$ with base point x so that I is an open interval and $f_{(I,\epsilon,i_\epsilon)}(y)$ is contained in $\overline{[\gamma]}$ for every $y \in I$. The union of the images of all such $f_{(I,\epsilon,i_\epsilon)}$ covers the rotational component except for possibly two linear approaches $[\gamma_{left}]$ and $[\gamma_{right}]$. In this case, half-closed intervals $I = (a, b]$ with $f_{(I,\epsilon,i_\epsilon)}(b) \in \{[\gamma_{left}], [\gamma_{right}]\}$ have to be allowed as open sets to obtain a cover of $\overline{[\gamma]}$.

We choose the collection of the corresponding sets $V(I, \epsilon, i_\epsilon)$ as a basis of the topology. In particular, any such $\varphi_{(I,\epsilon,i_\epsilon)}$ for I open is a homeomorphism.

Definition 1.19 (Translation structure on rotational components)
Let $(X, \mathcal{A})$ be a translation surface, $x \in \overline{X}$, and $\overline{[\gamma]}$ a rotational component of x that contains more than one linear approach. Then

$$\begin{aligned}\{(V(I,\epsilon,i_\epsilon),\varphi_{(I,\epsilon,i_\epsilon)}) :\ & (I,\epsilon,i_\epsilon) \text{ an angular sector with base point } x,\\ & I \text{ open},\ f_{(I,\epsilon,i_\epsilon)}(I) \subseteq \overline{[\gamma]}\}\\ \cup\ \{(V(I,\epsilon,i_\epsilon),\varphi_{(I,\epsilon,i_\epsilon)}) :\ & (I,\epsilon,i_\epsilon) \text{ an angular sector with base point } x,\\ & I = (a,b],\ f_{(I,\epsilon,i_\epsilon)}(I) \subseteq \overline{[\gamma]},\\ & f_{(I,\epsilon,i_\epsilon)}(b) \in \{[\gamma_{left}],[\gamma_{right}]\}\}\end{aligned}$$

forms an atlas of $\overline{[\gamma]}$.

Therefore, $\overline{[\gamma]}$ is a one-dimensional manifold, possibly with boundary. As the transition functions are actually translations in $\mathbb{R}$, $\overline{[\gamma]}$ even carries a one-dimensional translation structure.

For instance, the only rotational component of a cone angle singularity of multiplicity k is equipped with a one-dimensional translation structure so that it is isometric to $\mathbb{R}/2\pi k\mathbb{Z}$.

Using the one-dimensional translation structure on $\overline{[\gamma]}$, we can pull back the Euclidean metric from $\mathbb{R}$ to $\overline{[\gamma]}$ and for this we can measure the *length of a rotational component*.

We now describe the topology with which we endow the space of linear approaches and the space of rotational components.

Definition 1.20 (Topology on $\mathcal{L}(X)$ and $\mathcal{L}(x)$)
Let $(X, \mathcal{A})$ be a translation surface and $x \in \overline{X}$. For $\epsilon > 0$ we can define the *uniform metric* on $\mathcal{L}^\epsilon(X)$ using the translation metric d_X on X:

$$d_\epsilon(\gamma_1, \gamma_2) = \sup_{0<t<\epsilon} d_X(\gamma_1(t), \gamma_2(t)).$$

The uniform metric defines a topology on $\mathcal{L}^\epsilon(X)$. As $\mathcal{L}(X)$ is the colimit of the direct system of the spaces $\mathcal{L}^\epsilon(X)$, we can define the *final topology* on $\mathcal{L}(X)$. This is the finest topology so that all embeddings $\mathcal{L}^\epsilon(X) \hookrightarrow \mathcal{L}(X)$ are continuous.

The topology of $\mathcal{L}(x)$ is the induced topology as a subset of $\mathcal{L}(X)$.

The following characterization of the topology on $\mathcal{L}(X)$ from [BV13, Proposition 2.3] is more handsome than Definition 1.20.

Lemma 1.21 (Subbasis for the topology on $\mathcal{L}(X)$ [Bowman-Valdez]). Let $(X, \mathcal{A})$ be a translation surface. For every $x \in X$, $r > 0$, and $t > 0$, the set

$$\widetilde{B}(x,r)^t := \{[\gamma] : \gamma \in \mathcal{L}^\epsilon(X) \text{ for some } \epsilon > t,\ d(\gamma(t),x) < r\}$$

is an open set. In fact, the collection $\{\widetilde{B}(x,r)^t\}$ forms a subbasis for the topology on $\mathcal{L}(X)$ described in Definition 1.20.

The preceding lemma leads to a characterization of the topology on the space $\mathcal{L}(x)$ of linear approaches of $x \in X$ which will be often used in proofs and computations.

Lemma 1.22 (Subbasis for the topology on $\mathcal{L}(x)$). Let $(X, \mathcal{A})$ be a translation surface. For every geodesic curve $\gamma \in \mathcal{L}^\epsilon(x)$, $r > 0$, and $t < \epsilon$, the set

$$\begin{aligned} B(\gamma,t,r) &:= \widetilde{B}(\gamma(t),r)^t \cap \mathcal{L}(x) \\ &= \left\{[\gamma'] : \gamma' \in \mathcal{L}^{\epsilon'}(x) \text{ for some } \epsilon' > t,\ d(\gamma(t),\gamma'(t)) < r\right\} \end{aligned}$$

is an open neighborhood of $[\gamma]$ in $\mathcal{L}(x)$. In fact, the collection $\{B(\gamma,t,r)\}$ forms a subbasis for the topology on $\mathcal{L}(x)$ as a subspace of $\mathcal{L}(X)$.

Proof. The first statement is clear by the definition of the topology on $\mathcal{L}(x)$ and by $[\gamma] \in B(\gamma,t,r)$.

To prove the second statement, let $x' \in X$, $r' > 0$, and $t' > 0$. For every linear approach $[\gamma] \in \widetilde{B}(x',r')^{t'} \cap \mathcal{L}(x)$ we specify a neighborhood from that collection which is contained in $\widetilde{B}(x',r')^{t'} \cap \mathcal{L}(x)$: Let γ be a representative of $[\gamma]$ and define $t := t'$ and $r > 0$ small enough so that $B(\gamma(t),r) \subseteq B(x',r')$. Then we have

$$B(\gamma,t,r) \subseteq \widetilde{B}(x',r')^{t'} \cap \mathcal{L}(x)$$

and this shows that $\{B(\gamma,t,r)\}$ is a subbasis for the topology on $\mathcal{L}(x)$. □

For every $\epsilon > 0$ there exists a map $\operatorname{dir}_\epsilon \colon \mathcal{L}^\epsilon(X) \to S^1 = \mathbb{R}/2\pi\mathbb{Z}$ that associates to each geodesic curve in $\mathcal{L}^\epsilon(X)$ its direction. As all maps $\operatorname{dir}_\epsilon$ commute with the embeddings from the direct system and $\mathcal{L}(X)$ is the colimit of the direct system, we can define a map $\operatorname{dir} \colon \mathcal{L}(X) \to S^1$. Then we have the following useful lemma (see [BV13, Corollary 2.2]).

Lemma 1.23 (*Directions of linear approaches vary continuously* [*Bowman–Valdez*]). Let $(X, \mathcal{A})$ be a translation surface. Then $\operatorname{dir}\colon \mathcal{L}(X) \to S^1$ is a continuous map.

Now we can also define the space of rotational components as a quotient space of $\mathcal{L}(X)$.

Definition 1.24 (Space of rotational components)
Let $(X, \mathcal{A})$ be a translation surface and $x \in \overline{X}$. Recall that two linear approaches are called R-equivalent if they are contained in the same rotational component.

We define the *space of rotational components* as $\widetilde{\mathcal{L}}(X) := \mathcal{L}(X)/R$. It shall be endowed with the quotient topology. Furthermore, we define $\widetilde{\mathcal{L}}(x) := \mathcal{L}(x)/R$, also endowed with the quotient topology so that $\widetilde{\mathcal{L}}(x)$ is a subspace of $\widetilde{\mathcal{L}}(X)$.

Note that on each rotational component, we have a translation structure as defined in Definition 1.19 and we have on it the topology as a subspace of the space $\mathcal{L}(x)$ of linear approaches. These two topologies do not coincide in general as we will see explicitly in Section 2.3.

In Chapter 2, the spaces of linear approaches and the spaces of rotational components of some examples are studied in detail.

1.3 Ends of topological spaces

For connected, closed surfaces, i.e. connected, compact surfaces with empty boundary, we have the well-known result that the homeomorphism classes can be distinguished entirely by the genus and the information whether the surface is orientable. This result was first shown in special cases independently by Möbius in [Mö63] and by Jordan in [Jor66] and generally by von Dyck in [Dyc88] but all three mentioned proofs were not as rigorous as we would demand it today. A first rigorous proof was given by Dehn and Heegaard in [DH07] under the condition that the surfaces are triangulable. When combining this with Radó's result in [Rad25] that all surfaces are triangulable we have the following classification of closed surfaces (see [GX13, Appendix D] for more details on the history of the classification).

Theorem 1.25 (Classification of closed surfaces [Dehn-Heegaard, Radó])
Two connected, closed surfaces are homeomorphic if and only if they have the same genus and are in the same orientability class.

The genus of a surface is often defined by Betti numbers or by the idea of handles of a surface. Instead of that approach, we will employ a definition in Section 5.1 that uses the number of nonseparating curves (cf. [Zie08, Section 1.3]). For a connected surface X, a simple closed curve $\gamma\colon [0,l] \to X$ is called *nonseparating* if X with the image of γ removed is connected. In the same way, we say that a set of simple closed curves $\{\gamma_1,\ldots,\gamma_n\}$ is *nonseparating* if X with the images of $\gamma_1,\ldots,\gamma_n$ removed is connected.

Similarly, we say that a closed curve γ in $\overline{X}$ is *nonseparating* if $\overline{X} \setminus \mathrm{im}(\gamma)$ is connected and a set of closed curves $\{\gamma_1,\ldots,\gamma_n\}$ in $\overline{X}$ is *nonseparating* if $\overline{X} \setminus \mathrm{im}(\gamma_1) \cup \ldots \cup \mathrm{im}(\gamma_m)$ is connected.

Definition 1.26 (Genus)
Let X be a connected, orientable surface. The *genus* of X is $g \in \mathbb{N}$ if the following equivalent conditions are true.

(i) The maximum cardinality of a nonseparating set of disjoint curves in X is g.

(ii) The maximum cardinality of a nonseparating set of curves in X is $2g$.

Within the meaning of the previous definition, a connected, orientable surface X is said to have *infinite genus* if for every $n \in \mathbb{N} = \{0,1,\ldots\}$ there exists a nonseparating set of curves in X with cardinality n. This is equivalent to X containing subsurfaces of arbitrarily large genus.

As we are also interested in noncompact surfaces, possibly with infinite genus, we would like to have a similar classification result as in Theorem 1.25 for not necessarily compact surfaces. However, a cylinder of infinite length and the Euclidean plane are both orientable and have genus 0 but they are not homeomorphic. The Euclidean plane is homeomorphic to a once-punctured sphere whereas the cylinder is homeomorphic to a twice-punctured sphere. Informally, they are kept apart by the information that there are two possibilities to leave the surface for the cylinder and only one possibility for the plane.

A formalization of this concept of possibilities to leave the surface is the space of ends which was for the first time described by Kerékjártó in [Ker23, Abschnitt V, § 1]. The definition we give now goes back to an equivalent description by Hopf in [Hop43, § 1].

Definition 1.27 (Space of ends)
Let X be a path-connected, locally compact space.

(i) A *proper ray* in X is a continuous map $r\colon [0,\infty) \to X$ so that the preimage of every compact set in X is compact.

(ii) We say that two proper rays r_1, r_2 are *equivalent* if for every compact set $K \subseteq X$ there exists an $n \in \mathbb{N}$ so that $r_1([n,\infty))$ and $r_2([n,\infty))$ are contained in the same path-connected component of $X \setminus K$.

(iii) For a proper ray r, we call its equivalence class $\operatorname{end}(r)$ the *end* defined by r.

(iv) The set $\operatorname{Ends}(X) := \{\operatorname{end}(r) : r \text{ proper ray in } X\}$ is called *space of ends* of X.

The space of ends is empty for a compact space: As every proper ray has to leave any compact set, there can not exist proper rays into a compact space, so a compact space does not have ends. In contrast, $\mathbb{R}$ has two ends whereas $\mathbb{R}^n$ has only one end for $n \geq 2$.

We obtain by these definitions that every puncture of a surface gives rise to an end. Indeed, a proper ray defining this end could be a ray spiraling around the puncture, coming closer and closer to the puncture very slowly. We describe this construction more formally in Definition 3.2. In particular, our translation surfaces with cone angle singularities have ends. This kind of ends is sometimes excluded by other authors as the cone angle singularities are considered to be special points, included in the surface, in some definitions of translation surfaces.

The space of ends carries a topology that is defined by convergence: Let $r, r_1, r_2, \ldots$ be proper rays in a path-connected, locally compact space X. We say the sequence $\operatorname{end}(r_n)$ *converges* to $\operatorname{end}(r)$ for $n \to \infty$ if for every compact set $K \subseteq X$ there exists an $N_K \in \mathbb{N}$ and a sequence $(m_n)_{n \geq N_K} \subseteq \mathbb{N}$ so that $r_n([m_n,\infty))$ and $r([m_n,\infty))$ are contained in the same path-connected component of $X \setminus K$ for every $n \geq N_K$.

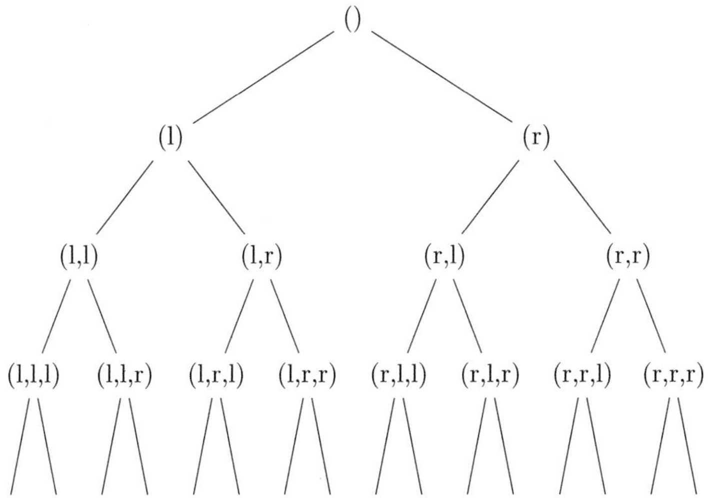

Figure 1.8: A vertex in the infinite complete binary tree can be described by a path from the root to the vertex, given by l(eft)'s and r(ight)'s.

As usual we say that a subset S of $\mathrm{Ends}(X)$ is *closed* if for every converging sequence of ends in S the limit is also contained in S. Note that the limit of a converging sequence is unique by the definition of the equivalence relation on the set of proper rays.

In the following more sophisticated example, we give an idea of what is possible for spaces of ends of path-connected, locally compact spaces.

Example 1.28 (*Space of ends of an infinite complete binary tree*). Let T be an infinite complete binary tree as in Figure 1.8. It has one vertex of degree 2 (called the root) and all other vertices have degree 3.

For every proper ray in T, there exists a unique equivalent ray that is geodesic and starts in the root. It can be described by an infinite sequence of r's and l's. Each two geodesic rays starting in the root differ after finite time, so they diverge and define different ends.

When replacing r and l by 0 and 2, we obtain the well-known description of the *Cantor (ternary) set* as the set of numbers in $[0,1] \subseteq \mathbb{R}$ whose ternary expansions in base 3 do not contain the digit 1.

Figure 1.9: The two surfaces each have infinite genus and two ends but they are not homeomorphic to each other.

The topology of the Cantor ternary set as a subset of $\mathbb{R}$ is the subspace topology. In Ends(T), for the end of a geodesic ray starting in the root, a neighborhood subbasis of the topology is given by the ends of all geodesic rays starting in the root for which the sequence of r's and l's agrees with the sequence of the given ray in the first n elements for a fixed $n \in \mathbb{N}$. Hence, there exists a homeomorphism between the ends of T (which are the ends defined by geodesic rays starting in the root) and the Cantor ternary set.

As the example of the infinite complete binary tree indicates, the space of ends of a path-connected, locally compact space does not have to be finite. However, the space of ends has some remarkable properties as was shown in [Fre31, Satz 4 and Satz 7].

Proposition 1.29 (Properties of the space of ends [Freudenthal]) *The space of ends of a path-connected, locally compact space is totally disconnected, separable, and compact. In particular, it is homeomorphic to a closed subset of the Cantor set.*

We defined the space of ends to distinguish between different homeomorphism classes of surfaces. However, the two surfaces in Figure 1.9 are not homeomorphic to each other although they have the same genus and the space of ends is in both cases the discrete two-point space. Though, in a visual way we can distinguish two kinds of ends in the examples in Figure 1.9: one kind with genus, one without. This leads us to the following definition.

Definition 1.30 (Planar and orientable ends)
Let X be a connected surface.

(i) An end e of X is called *orientable* if there exist a proper ray r with $\text{end}(r) = e$, a compact set $K \subseteq X$, and an $n \in \mathbb{N}$ so that the path-connected component of $X \setminus K$, in which $r([n, \infty))$ is contained, is orientable.

Otherwise, the end e is called *nonorientable.*

(ii) An end e of X is called *planar* if there exist a proper ray r with $\text{end}(r) = e$, a compact set $K \subseteq X$, and an $n \in \mathbb{N}$ so that the path-connected component of $X \setminus K$, in which $r([n, \infty))$ is contained, is planar, i.e. orientable and of genus 0.

Otherwise, the end e is called *nonplanar.*

(iii) The spaces $\text{Ends}_{nonorientable}(X)$ and $\text{Ends}_{nonplanar}(X)$ are defined as the subspaces of $\text{Ends}(X)$ that contain exactly the nonorientable and nonplanar ends, respectively.

If an orientable surface has finite genus then all of its ends have to be planar. If the surface has infinite genus then at least one of the ends is nonplanar.

In Figure 1.9, the upper surface has one planar end and one nonplanar but orientable end. The lower surface has two nonplanar but orientable ends. It can be shown that a homeomorphism can not map a planar end to a nonplanar end. Hence, the properties of ends from Definition 1.30 give us the necessary distinction to express that the two surfaces in Figure 1.9 are not homeomorphic.

The general statement of classifying surfaces by their space of ends was first proven by Kerékjártó in [Ker23, Abschnitt V, § 1]. A more modern proof is given in [Ric63] by Richards who also introduces four *orientability classes* in the case of noncompact surfaces: orientable, infinitely nonorientable, odd nonorientable, and even nonorientable. As we consider only orientable surfaces, we will not discuss the details of orientability classes here.

Theorem 1.31 (Classification of surfaces [Kerékjártó])
Let X, Y be two connected surfaces of the same genus and the same

orientability class. Then X and Y are homeomorphic if and only if the nested triples of spaces

$$\mathrm{Ends}(X) \supseteq \mathrm{Ends}_{nonplanar}(X) \supseteq \mathrm{Ends}_{nonorientable}(X)$$

and

$$\mathrm{Ends}(Y) \supseteq \mathrm{Ends}_{nonplanar}(Y) \supseteq \mathrm{Ends}_{nonorientable}(Y)$$

are homeomorphic.

As translation surfaces are always orientable, we can describe their homeomorphism classes completely by the genus, the space of ends, and the space of nonplanar ends. Some common examples of equivalence classes of orientable surfaces were given names: the surfaces described in (iv) and (v) of the following example by Phillips and Sullivan in [PS81, Section 1] and the surfaces described in (iii) and (vi) by Ghys in [Ghy95, Théorème A].

Example 1.32 (*Surfaces of genus* 0 *or* ∞ *with one, two, or infinitely many ends*).

(i) A surface of genus 0 with one end is homeomorphic to $\mathbb{R}^2$.

(ii) A surface of genus 0 with two ends is homeomorphic to a cylinder of infinite length.

(iii) The surface of genus 0 with uncountably many ends shown in Figure 1.10 is called *Cantor tree*. Be aware that this surface still has a countable basis of topology in spite of the uncountably many ends.

(iv) A surface of infinite genus with one end (which is automatically nonplanar) is called *Loch Ness monster*.

(v) A surface of infinite genus with two ends that are both nonplanar is called *Jacob's ladder*.

(vi) The surface of infinite genus with uncountably many ends that are all nonplanar shown in Figure 1.10 is called *blooming Cantor tree*.

Most interesting examples of wild translation surfaces with one singularity are Loch Ness monsters, for example the Chamanara surface (see Section 2.1).

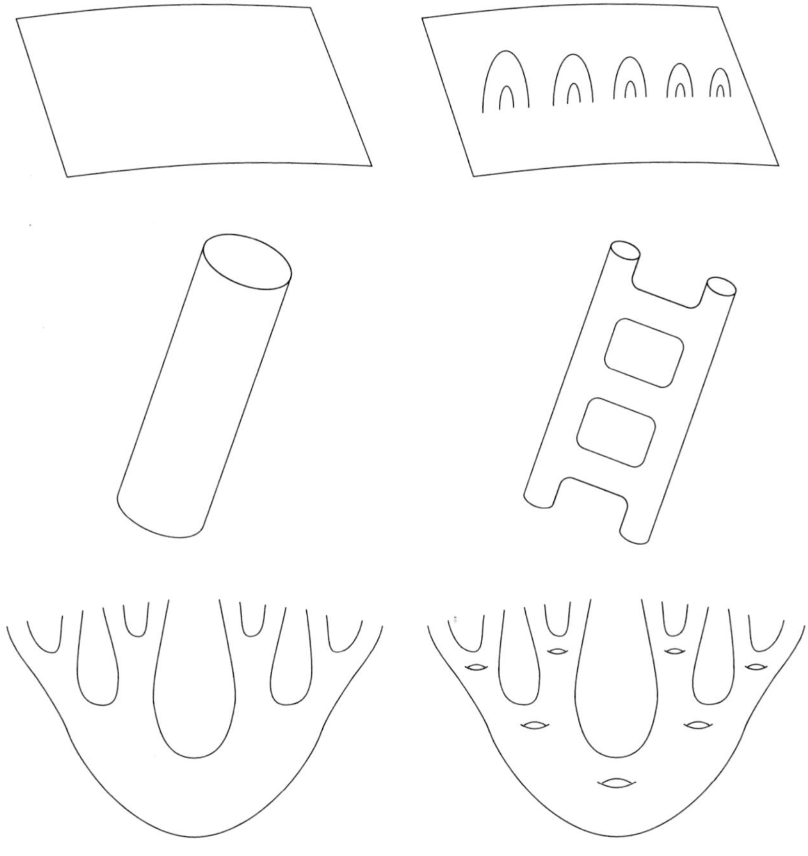

Figure 1.10: Sketches of $\mathbb{R}^2$, Loch Ness monster, cylinder, Jacob's ladder, Cantor tree, and blooming Cantor tree (from top left to bottom right).

1.4 Veech groups of translation surfaces

The first study of Veech groups was done in [Vee89] by Veech after whom the groups are named. He defined them using affine maps to measure the symmetry of a finite translation surface. The Veech group of a finite translation surface contains information about the dynamical properties of the surface. A nice collection of fundamental results on Veech groups can be found in [HS06].

A continuous map $f\colon (X,\mathcal{A}) \to (Y,\mathcal{B})$ of translation surfaces is called *affine* if it is locally affine, i.e. for every $x \in X$ there exist charts $(U,\varphi) \in \mathcal{A}$ and $(V,\psi) \in \mathcal{B}$ with $x \in U$ and $f(U) \subseteq V$ so that for every $z \in \varphi(U) \subseteq \mathbb{R}^2$ it is true that

$$\left(\psi \circ f \circ \varphi^{-1}\right)(z) = A \cdot z + t \text{ for an } A \in \mathrm{GL}(2,\mathbb{R}) \text{ and a } t \in \mathbb{R}^2.$$

The matrix A is globally the same for an affine map f and it is called the *derivative* of f. An affine map is called *translation* if its derivative is the identity matrix.

Definition 1.33 (Affine group and Veech group)
Let $(X,\mathcal{A})$ be a translation surface.

(i) The *affine group* $\mathrm{Aff}^+(X,\mathcal{A})$ of $(X,\mathcal{A})$ is defined as

$$\mathrm{Aff}^+(X,\mathcal{A}) := \big\{f\colon X \to X : f \text{ is an orientation-preserving homeomorphism and affine with respect to } \mathcal{A}\big\}.$$

(ii) The map $\mathrm{der}\colon \mathrm{Aff}^+(X,\mathcal{A}) \to \mathrm{GL}^+(2,\mathbb{R})$ that assigns to each affine map its derivative is called *derivation map*.

(iii) The image of der is called *Veech group* of $(X,\mathcal{A})$ and denoted by $\mathrm{GL}^+(X,\mathcal{A})$.

For a translation surface with finite area, the elements of the affine group are area-preserving, hence the Veech group is a subgroup of $\mathrm{SL}(2,\mathbb{R})$. For a finite translation surface, it is even a discrete subgroup of $\mathrm{SL}(2,\mathbb{R})$ as can be shown by using the discreteness of the set of holonomy vectors of saddle connections (recall the remark after Definition 1.11 or see [Vee89, Proposition 2.7] for the original proof).

For the Euclidean plane as a translation surface, the Veech group is $\mathrm{GL}^+(2,\mathbb{R})$. The nontrivial standard example of the calculation of a Veech group is the one of a torus.

Example 1.34 (Veech group of a torus and a cylinder).

(i) Let $(T,\mathcal{A})$ be a torus so that there exist $a,b,c,d \in \mathbb{R}$ with

$$B := \begin{pmatrix} a & b \\ c & d \end{pmatrix} \in \mathrm{GL}(2,\mathbb{R})$$

and $T = \mathbb{C}/\Lambda_B$ for the lattice $\Lambda_B := (a+\mathrm{i}c)\mathbb{Z} + (b+\mathrm{i}d)\mathbb{Z}$.

Every element in $\mathrm{Aff}^+(T,\mathcal{A})$ corresponds to an element of the affine group of the Euclidean plane which preserves the lattice Λ_B. Therefore, a matrix $A \in \mathrm{GL}^+(2,\mathbb{R})$ is the derivative of an element in $\mathrm{Aff}^+(T,\mathcal{A})$ if and only if we have that for all $n,m \in \mathbb{Z}$ there exist $n',m' \in \mathbb{Z}$ with

$$A \cdot B \cdot \begin{pmatrix} n \\ m \end{pmatrix} = B \cdot \begin{pmatrix} n' \\ m' \end{pmatrix}.$$

This is equivalent to $B^{-1}AB \in \mathrm{GL}(2,\mathbb{Z})$ and as A has positive determinant, we even have $B^{-1}AB \in \mathrm{SL}(2,\mathbb{Z})$. Hence, we obtain

$$\mathrm{GL}^+(T,\mathcal{A}) = B \cdot \mathrm{SL}(2,\mathbb{Z}) \cdot B^{-1}.$$

(ii) Let $(C,\mathcal{A})$ be a cylinder so that there exists $v = v_1 + \mathrm{i}v_2 \in \mathbb{C}$ with $C = \mathbb{C}/v\mathbb{Z}$. Again, an affine map f is contained in $\mathrm{Aff}^+(C,\mathcal{A})$ if and only if $\mathrm{der}(f) = A \in \mathrm{GL}^+(2,\mathbb{R})$ and f preserves $v\mathbb{Z}$. Therefore, for $A \in \mathrm{GL}^+(C,\mathcal{A})$ there exists an $n \in \mathbb{Z}$ with

$$A \cdot \begin{pmatrix} v_1 \\ v_2 \end{pmatrix} = n \cdot \begin{pmatrix} v_1 \\ v_2 \end{pmatrix}.$$

As this is also true for A^{-1}, it follows that n is an invertible integer, hence 1 or -1.

In the special case of a vertical cylinder, we have $v_2 = 0$. It follows that the only condition on a Veech group element is that its first column is $\pm\left(\begin{smallmatrix} 1 \\ 0 \end{smallmatrix}\right)$. From this it follows for a vertical cylinder $(C,\mathcal{A})$:

$$\mathrm{GL}^+(C,\mathcal{A}) = \left\{\begin{pmatrix}1 & t\\ 0 & s\end{pmatrix} : t\in\mathbb{R}, s\in\mathbb{R}_+\right\} \cup \left\{\begin{pmatrix}-1 & t\\ 0 & -s\end{pmatrix} : t\in\mathbb{R}, s\in\mathbb{R}_+\right\}.$$

The action of the affine group on a translation surface $(X,\mathcal{A})$ preserves singularities and the property of a segment to be geodesic. Hence, for every $f \in \mathrm{Aff}^+(X,\mathcal{A})$ we can define $f_\epsilon\colon \mathcal{L}^\epsilon(X) \hookrightarrow \mathcal{L}(X)$ by sending a geodesic curve $\gamma\colon (0,\epsilon) \to X$ to the linear approach $[f\circ\gamma]$. As all maps f_ϵ commute with the embeddings from the direct system, we can define a map $f_*\colon \mathcal{L}(X) \to \mathcal{L}(X)$. Then we have the following statement (see [BV13, Theorem 3.2]).

Proposition 1.35 (From $\mathrm{Aff}^+(X,\mathcal{A})$ to $\mathrm{Homeo}(\mathcal{L}(X))$ [Bowman–Valdez])
Let $(X,\mathcal{A})$ be a translation surface and $f \in \mathrm{Aff}^+(X,\mathcal{A})$ an affine homeomorphism. Then $f_*\colon \mathcal{L}(X) \to \mathcal{L}(X)$ is a homeomorphism.

Recall that $\mathrm{SL}(2,\mathbb{R})$ and $\mathrm{PSL}(2,\mathbb{R}) := \mathrm{SL}(2,\mathbb{R})/\{\pm I\}$ act transitively on the upper half-plane $\mathbb{H}$ and more generally on the Riemann sphere $\mathbb{C}\cup\{\infty\}$ by *Möbius transformations*, defined as

$$A \bullet z := \frac{az+b}{cz+d} \text{ for } z\in\mathbb{C} \text{ and } A = \begin{pmatrix}a & b\\ c & d\end{pmatrix} \in \mathrm{SL}(2,\mathbb{R}).$$

The elements of $\mathrm{SL}(2,\mathbb{R})$ or $\mathrm{PSL}(2,\mathbb{R})$ except for (the negative of) the identity matrix can be divided into the three classes of parabolic, elliptic, and hyperbolic elements. The classification can be done by the traces of the matrices, by their fixed points, or by conjugacy classes (see Table 1.1).

The parabolic elements in the Veech group are closely related to shears along *cylinders* in the translation surface.

Definition 1.36 (Cylinder and cylinder decomposition)
Let $(X,\mathcal{A})$ be a translation surface with at least one singularity.

(i) A *cylinder* in $(X,\mathcal{A})$ of *circumference* $k>0$ and *height* $a>0$ is an open subset of X which is isometric to a Euclidean cylinder $\mathbb{R}/k\mathbb{Z}\times(0,a)$.

	absolute value of trace of the element	fixed points of the action on $\mathbb{C} \cup \infty$	conjugacy classes
elliptic	less than 2	two fixed points in $\mathbb{C} \setminus \mathbb{R}$, conjugated to each other	conjugated to a rotation, that means to $\begin{pmatrix} \cos\theta & \sin\theta \\ -\sin\theta & \cos\theta \end{pmatrix}$ for a $\theta \in (0, 2\pi)$
parabolic	equal to 2	one fixed point in $\mathbb{R} \cup \{\infty\}$	conjugated to a shear, that means to $\begin{pmatrix} 1 & \pm 1 \\ 0 & 1 \end{pmatrix}$ or to $\begin{pmatrix} -1 & \pm 1 \\ 0 & -1 \end{pmatrix}$
hyperbolic	greater than 2	two fixed points in $\mathbb{R} \cup \{\infty\}$	conjugated to a squeeze mapping, that means to $\begin{pmatrix} \lambda & 0 \\ 0 & \frac{1}{\lambda} \end{pmatrix}$ for a $\lambda \in \mathbb{R}$ with $\lvert\lambda\rvert > 1$

Table 1.1: Characterization of elliptic, parabolic, and hyperbolic elements in $\mathrm{SL}(2, \mathbb{R})$ by different aspects.

(ii) The *modulus* m of a cylinder is the ratio of circumference and height, i.e. $m = \frac{k}{a}$.

(iii) If a cylinder can be extended to a *maximal cylinder* then the maximal cylinder is bounded by saddle connections. The *direction* of the cylinder is the direction of the saddle connections.

(iv) A *cylinder decomposition* of $(X, \mathcal{A})$ is a collection of maximal cylinders in $(X, \mathcal{A})$ so that the closures of the cylinders in X cover X and so that each two cylinders are disjoint. The *direction* of the cylinder decomposition is the direction of the cylinders.

Note that the union of the closures of the cylinders does not need to be the same as the closure of the union of the cylinders. We will see in Section 2.2 what this distinction implies for the existence of a cylinder decomposition in a given direction.

To illustrate the concept of cylinder decompositions, we give a classical example.

Example 1.37 (Cylinder decomposition of Veech's double n-gons). Recall the double n-gon from Example 1.7 for $n = 5$. This translation surface has a horizontal cylinder decomposition as indicated in Figure 1.11.

The striped cylinder has height $\cos \frac{3\pi}{10} = \sin \frac{\pi}{5}$ and circumference $2 \sin \frac{3\pi}{10} = 2 \cos \frac{\pi}{5}$, hence the modulus is $2 \cot \frac{\pi}{5}$. The dotted cylinder has height $\cos \frac{\pi}{10} = 2 \cdot \sin \frac{3\pi}{10} \cdot \sin \frac{\pi}{5}$ and circumference $2 + 2 \sin \frac{\pi}{10} = 4 \cdot \sin \frac{3\pi}{10} \cdot \cos \frac{\pi}{5}$, hence the same modulus $2 \cot \frac{\pi}{5}$.

It is not a special feature of the double pentagon to have a cylinder decomposition such that all cylinders have the same modulus. In fact,

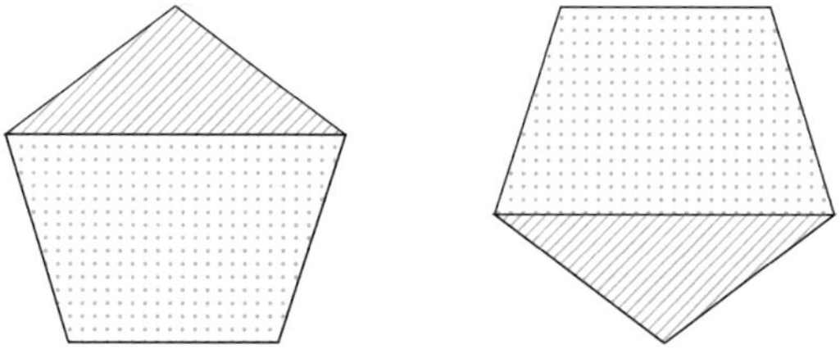

Figure 1.11: Cylinder decomposition of Veech's double n-gon for $n = 5$.

Veech calculated in [Vee89, Section 5] that such a cylinder decomposition exists for all double n-gons. This makes it easy to detect elements of the Veech group as we will see in Proposition 1.38.

Also the infinite staircase in Example 1.8 and the Chamanara surface in Example 1.9 have cylinder decompositions. We justify the latter statement in more detail in Section 2.1, in the context of the interaction with parabolic elements in the Veech group.

Proposition 1.38 (Cylinder decompositions and parabolic elements [Veech])

(i) Let $(X, \mathcal{A})$ be a translation surface and (z_n) a cylinder decomposition so that the cylinder z_n has height h_n and circumference c_n. If the inverse moduli $\frac{h_n}{c_n}$ are commensurable, *i.e. if there exists an $m \in \mathbb{R}$ so that each inverse modulus is an integer multiple of m, then the Veech group contains a matrix conjugated to $\begin{pmatrix} 1 & \frac{1}{m} \\ 0 & 1 \end{pmatrix}$.*

(ii) Let $(X, \mathcal{A})$ be a finite translation surface such that the Veech group contains a parabolic element. Then there exists a cylinder decomposition of $(X, \mathcal{A})$ in the eigen direction of the parabolic element.

Proof. The statement is classical for finite translation surfaces and was first used in [Vee89]. The proof of (i) given there literally works for the general case. We refer to [HS06, Lemma 4] in conjunction with [MT02, Theorem 1.8] for a more self-contained proof of (ii). □

For finite translation surfaces, every cylinder composition can only contain finitely many cylinders. So in this case, the number $\frac{1}{m}$ can be chosen to be the least common multiple of the moduli.

1.5 Dynamics on translation surfaces

The main object in the theory of dynamics on translation surfaces is a *flow*. We will define flows and some properties related to flows and state a theorem of Poincaré which will be helpful in Section 5.2.

Definition 1.39 (Flow)
A *flow* on a space X is an action of $\mathbb{R}$ on X.

Most authors want flows to preserve the structure of X, for example flows on topological spaces shall be continuous and flows on differentiable manifolds shall be differentiable. We do not need to discuss this further as we are only interested in a special example of a flow – in the *geodesic flow* on a translation surface $(X, \mathcal{A})$.

To be precise, the geodesic flow on $(X, \mathcal{A})$ is not a flow according to Definition 1.39. It is an action of $\mathbb{R}$ on the unit tangent bundle $T^1(X)$ of $(X, \mathcal{A})$, equipped with the natural flat metric, which is not defined for all pairs in $T^1(X) \times \mathbb{R}$: Let $(x, v) \in T^1(X) = X \times S^1$ and $r \in \mathbb{R}$. If there exists a geodesic starting in x and with holonomy vector of length r and direction v then it is unique and we define $x' \in X$ to be the end point of the geodesic. Furthermore, we define $F((x, v), r) = (x', v)$ if x' is defined. The definition indicates that it can happen that $\lim_{r \to r_0} F((x, v), r)$ should be a point in the tangent space at a singularity for a fixed r_0. In this case, the geodesic flow F is not defined for (x, v) and $r \geq r_0$. However, this is the only obstacle for the flow to be not defined.

By a geodesic flow in a given direction $\theta \in S^1$, we mean the restriction of F to all points in the unit tangent bundle with direction θ. It will be denoted by $F_\theta \colon X \times \mathbb{R} \to X$ but again F_θ is not defined on all elements in $X \times \mathbb{R}$ so it is not a flow according to Definition 1.39.

The orbit of a point under a flow is called *trajectory*. The deficiency of a geodesic flow F_θ to be a flow is that there exist trajectories that are not defined for every $r \in \mathbb{R}$ if the translation surface has singularities.

Definition 1.40 (Periodic and recurrent flow)
Let $(X, \mathcal{A})$ be a translation surface, $\theta \in S^1$, and F_θ the geodesic flow on $(X, \mathcal{A})$ in direction θ.

(i) The geodesic flow F_θ is called *periodic* if all trajectories that are defined for all time, i.e. for every $r \in \mathbb{R}$, are closed.

(ii) A point $x \in X$ is called *recurrent* under F_θ if for every neighborhood U of x and every $r_0 \in \mathbb{R}$ there exists an $r > r_0$ with $F_\theta(x, r) \in U$.

(iii) The geodesic flow F_θ is called *recurrent* if almost every point with respect to the Lebesgue measure on X is recurrent under F_θ.

The following statement was shown in [Poi90, Théorème I].

Proposition 1.41 (Poincaré recurrence [Poincaré])
Let X be a finite measure space and $F\colon X \to X$ a measure-preserving transformation. Then F is recurrent.

When we consider a translation surface of finite area and let X be the set of all regular points in the translation surface for which the geodesic flow is defined for all time, then Poincaré recurrence is applicable. Note that for a feasible statement it is crucial that the flow is defined for all time on almost all points. We will come back to this condition in Section 5.2.

It has to be said that the theory of dynamics on translation surfaces is much richer than the extract which is presented here. For finite translation surfaces, the geodesic flow on a given translation surface is closely related to the Teichmüller flow on the moduli space of all finite translation surfaces with the same genus. This relation gives the possibility to produce many deep results like the Veech dichotomy in [Vee89] and the famous recent results of Eskin, Mirzakhani, and Mohammadi in [EMM15] and of Eskin and Mirzakhani in [EM13].

However, for translation surfaces in general, the discussion on a good definition of moduli spaces has just started (see [Hoo13b] and [Hoo13a] for a first approach), so there is no relation established as for the finite translation surfaces. We nevertheless introduce the language of dynamics here to clearly state some conditions on translation surfaces in Section 5.2.

2 Examples that are worth knowing

In this chapter, we give three examples with various remarkable properties and a construction to produce many more examples. In particular, the singularities and their spaces of linear approaches and the homeomorphism class of the surface are determined for each example.

2.1 Chamanara surface

History In [Cha04], Chamanara describes in detail a family of translation surfaces with a parameter $\alpha \in (0,1)$. For $\alpha = \frac{1}{2}$, the translation surface from [Cha04] is exactly the one described in Example 1.9. In general, the square has edges of length $\sum_{i=1}^{\infty} \alpha^n$ for an $\alpha \in (0,1)$ and the edges are divided into segments where the nth segment has length α^n.

We will consider the translation surface with parameter $\alpha = \frac{1}{2}$ as *the* Chamanara surface in this section and throughout the thesis.

Veech group The Veech group is generated by the two parabolic elements

$$\begin{pmatrix} -2 & 3 \\ -3 & 4 \end{pmatrix}, \begin{pmatrix} -2 & \frac{3}{2} \\ -6 & 4 \end{pmatrix}$$

and $-\mathrm{Id}$ (see [Cha04, Theorem 4]). In the proof, the existence of parabolic elements in the Veech group is shown by the existence of cylinders with commensurable inverse moduli (see Proposition 1.38). We need some knowledge on these cylinders later so we describe them briefly here.

Figure 2.1 represents a cylinder decomposition of slope 4 on the Chamanara surface. There are infinitely many cylinders, all except one built by two trapezoids. As the trapezoids are similar, the moduli of these cylinders coincide. In fact, this modulus can be calculated to be $\frac{4}{51}$ by

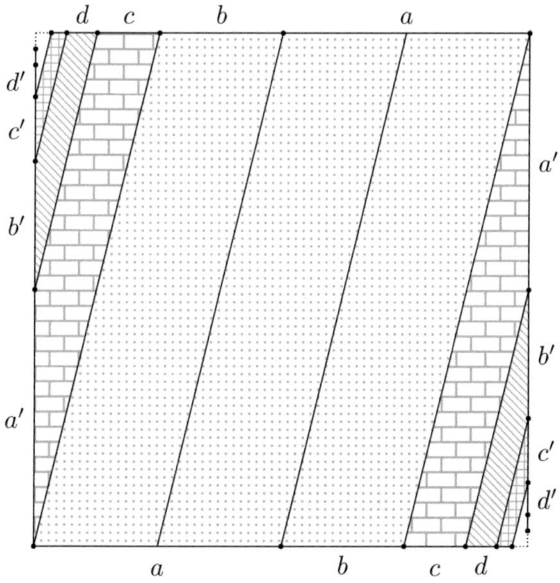

Figure 2.1: Chamanara surface with a cylinder decomposition of slope 4.

trigonometry. Similar calculations show that the modulus of the middle cylinder is $\frac{4}{51}$, too. Then Proposition 1.38 implies that there exists a parabolic element in the Veech group which maps every saddle connection of slope 4 to itself and acts as a Dehn twist on every cylinder in the cylinder decomposition.

For every slope 2^n with $n \in \mathbb{Z}$ there exists a cylinder decomposition like the previous one (see [Cha04, Proposition 11]). We have that every maximal geodesic of slope 2^n starting in a singularity is a saddle connection and bounds a cylinder of the same slope. As in the previous case, we have infinitely many cylinders built by two trapezoids and finitely many cylinders built by parallelograms like the middle cylinder in the previous case. Again, the moduli of the cylinders of the first type are all the same. Let h be the height of the largest cylinder of the first type and c its circumference (see Figure 2.2). Then we have for a cylinder of the second type that its height is an integer multiple of h and its circumference is

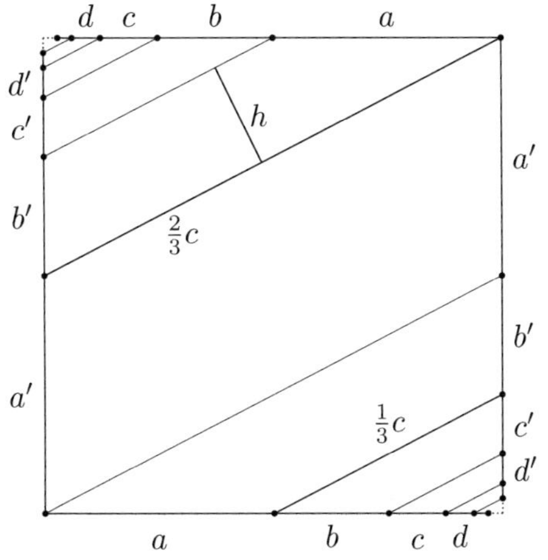

Figure 2.2: Chamanara surface with saddle connections of slope $\frac{1}{2}$: h is the height and c is the circumference of the largest cylinder that is built by two trapezoids.

an integer multiple of $\frac{2}{3}c$. Therefore the inverse modulus of this cylinder is a rational multiple of $\frac{3}{2} \cdot \frac{h}{c}$. For example, for slope 2 and slope $\frac{1}{2}$, the inverse modulus of the middle cylinder is three times the other moduli. This yields the existence of a number m such that for every cylinder C_i there exists an integer k_i such that the inverse modulus of C_i is $k_i \cdot m$. In this case, there exists a parabolic element in the Veech group such that the corresponding affine map twists the ith cylinder k_i times.

Spaces of linear approaches and rotational components We showed in Example 1.15 that the Chamanara surface has exactly one singularity σ. The next task is to understand the space $\mathcal{L}(\sigma)$ of linear approaches and the space $\widetilde{\mathcal{L}}(\sigma)$ of rotational components. In [BV13, Example 1.10], Bowman and Valdez show that there exist two rotational components of infinite length and infinitely many rotational components of finite length. We follow that discussion and add statements on the topology of the space of linear approaches and the space of rotational components. The main objective is to obtain a characterization for open sets in $\widetilde{\mathcal{L}}(\sigma)$: a nontrivial set in $\widetilde{\mathcal{L}}(\sigma)$ is open if and only if it consists of a single rotational component of infinite length or if it contains both rotational components of infinite length. This was done with Lucien Clavier and Chenxi Wu and previously published in [CRW14, Example 2.4].

Consider one of the points on the edges of the square where a segment is halved and consider a direction going inwards from that cutting point. Then there exists a linear approach in that direction with base point the cutting point. Also, for the corners of the square, we find linear approaches. These linear approaches give us four rotational components, two of infinite length and isometric to $\mathbb{R}$ (see Figure 1.6 again) and two isometric to $(0, \frac{\pi}{2})$ (in the upper left and the lower right corner). The obstacles that prevent the rotational components of finite length from being extended further are the cutting points which are coming closer to the corners. However, there are more linear approaches which are not readily visible in the figure.

Consider a linear approach in one of the two rotational components of length $\frac{\pi}{2}$. For a cylinder decomposition as discussed above, each representative of the linear approach crosses infinitely many cylinders. Now we apply the affine map to the Chamanara surface which is a (possibly multiple) Dehn twist on each cylinder of the chosen cylinder decomposition. Then the representative of the linear approach is mapped to a geodesic curve of finite length, starting in a singularity. Hence, the image is again a representative of a linear approach. The corresponding rotational component is the image of the original rotational component of length $\frac{\pi}{2}$ under the homeomorphism of $\mathcal{L}(\sigma)$ which is induced by the (multiple) Dehn twist (cf. Proposition 1.35). Observe that a linear approach in such a rotational component intersects the edges of the square infinitely often.

Lemma 2.1 (Directions occur at most once in rotational components of finite length). Let $(X, \mathcal{A})$ be the Chamanara surface, σ the unique singularity, and $[\gamma] \in \mathcal{L}(\sigma)$ a linear approach contained in a rotational component of finite length. Then every other linear approach in $\mathcal{L}(\sigma)$ with the same direction is contained in a rotational component of infinite length.

Proof. Note that all linear approaches with base point one of the cutting points or with base point the lower left or upper right corner are contained in rotational components of infinite length. Hence, we are only interested in linear approaches with base point the lower right or the upper left corner.

Let $\theta \in S^1 = \mathbb{R}/2\pi\mathbb{Z}$ be the direction of $[\gamma]$. Because of the symmetry and because of the cutting points close to the lower right and upper left corner, we can restrict to $\theta \in (0, \pi)$. We first suppose $\theta \in (\frac{\pi}{2}, \pi)$. Then $[\gamma]$ is the unique linear approach with base point the lower right corner and with direction θ. A linear approach with base point the upper left corner and direction θ does not exist.

Now let $\theta \in (0, \frac{\pi}{2})$. Note that θ can not correspond to the slope of a cylinder decomposition as $[\gamma]$ would then define a saddle connection and would have a cutting point as base point. Hence, let $n \in \mathbb{Z}$ be a number for which the slope 2^n is close to the direction θ and so that every (short enough) representative γ of $[\gamma]$ traverses the cylinders of the cylinder decomposition of slope 2^n one by one and γ runs along the core curve of each cylinder of the cylinder decomposition the same number of times. Let this number be $m \in \mathbb{Z}$.

Let $f\colon X \to X$ be the affine map which twists the cylinders of the cylinder decomposition of slope 2^n as in the previous discussion and let $f_*\colon \mathcal{L}(\sigma) \to \mathcal{L}(\sigma)$ be the induced homeomorphism on the linear approaches. Then f^m maps γ to a geodesic curve which traverses each cylinder straightly without running along the core curve (possibly after applying this process more than one time).

This implies that $f^m(\gamma)$ has direction $\theta' \in (\frac{\pi}{2}, \pi) \cup (\frac{3\pi}{2}, 0)$. Hence, $f_*([\gamma])$ is the unique linear approach with base point the lower right corner or upper left corner and with direction θ'. In particular, $[\gamma]$ is the unique linear approach with base point the lower right or upper left corner and with direction θ. □

Proposition 2.2 (Topology of $\widetilde{\mathcal{L}}(\sigma)$)
Let $(X, \mathcal{A})$ be the Chamanara surface and σ the unique singularity. A set in $\widetilde{\mathcal{L}}(\sigma)$ is open if and only if it is empty, if it consists of a single rotational component of infinite length, or if it contains both rotational components of infinite length.

Proof. By Lemma 2.1 together with Lemma 1.23 we can deduce that the subspace of all linear approaches contained in a fixed rotational component of finite length has a neighborhood which does not contain any linear approach of another rotational component of finite length.

On the other hand, such a neighborhood contains linear approaches from both of the rotational components of infinite length as we will show

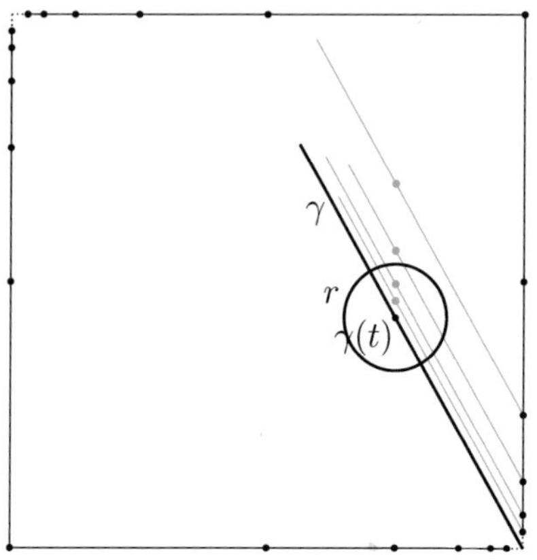

Figure 2.3: Every neighborhood of the linear approach $[\gamma]$ contains linear approaches from both rotational components of infinite length.

now. Recall from Lemma 1.22 that for a linear approach $[\gamma] \in \mathcal{L}(\sigma)$ with representative γ and $t > 0$, $r > 0$ the set

$$B(\gamma, t, r) = \{[\gamma'] : \gamma' \in \mathcal{L}^{\epsilon}(\sigma) \text{ for some } \epsilon > t,\ d(\gamma(t), \gamma'(t)) < r\} \subseteq \mathcal{L}(\sigma)$$

is an open neighborhood of $[\gamma]$ in $\mathcal{L}(\sigma)$. Then there exists a linear approach in one of the rotational components of infinite length that is contained in $B(\gamma, t, r)$ as indicated in Figure 2.3. In particular, for such a linear approach $[\gamma]$, there exists a sequence of linear approaches in each rotational component of infinite length converging to $[\gamma]$.

In contrary, for every linear approach $[\gamma]$ in a rotational component of infinite length, there exists an open interval I and an angular sector $(I, \epsilon, i_\epsilon)$ such that the image of $f_{(I,\epsilon,i_\epsilon)}$ is an open neighborhood of $[\gamma]$. However, every linear approach in the image of $f_{(I,\epsilon,i_\epsilon)}$ is contained in the rotational component of $[\gamma]$ by the definition of rotational component. Hence, this neighborhood does not contain any linear approach of another rotational component.

In summary we have that the rotational components of infinite length are open points in $\widetilde{\mathcal{L}}(\sigma)$ whereas each neighborhood of a rotational component of finite length contains both rotational components of infinite length. $\square$

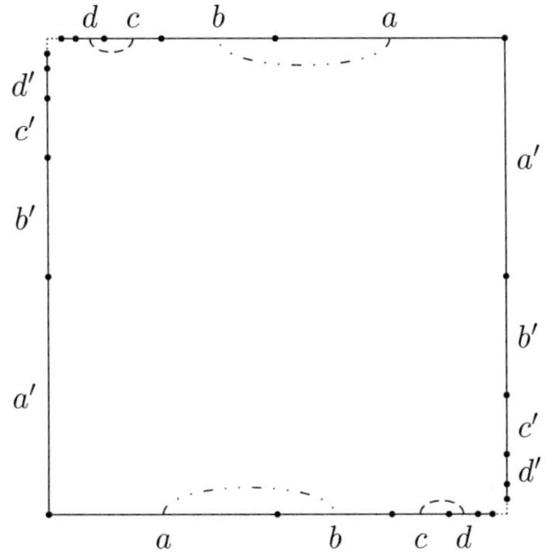

Figure 2.4: Taking out the images of all curves of this type does not separate the Chamanara surface.

Topology of the surface Following Definition 1.26, we can show infinite genus by specifying a nonseparating set of curves of infinite cardinality. For the Chamanara surface we can do this as in Figure 2.4: the nth curve connects the middle of the $(2n-1)$th segment on the top to the middle of the $2n$th segment on the top (which is identified with the $2n$th segment on the bottom) and back to the middle of the $(2n-1)$th segment on the bottom (which is identified with the $(2n-1)$th segment on the top). This set is nonseparating since every point in the complement of the images of the curves can be connected to the center of the square by a curve in the complement, possibly by first crossing an edge of the square.

To specify the homeomorphism class of the Chamanara surface, we also have to determine the space of ends of X. For this, let $K \subseteq X$ be a compact set and $\epsilon := d(K, \sigma)$. As $K \subseteq X$ is compact, ϵ is positive and hence K is contained in the compact set $X \setminus B(\sigma, \epsilon)$. Furthermore, let $r_1, r_2\colon [0, \infty) \to X$ be two proper rays. Because of the properness there exists an $n \in \mathbb{N}$ such that $r_1([n, \infty))$ and $r_2([n, \infty))$ are contained in $B(\sigma, \epsilon) \setminus \{\sigma\}$. The set $B(\sigma, \epsilon) \setminus \{\sigma\} \subseteq X \setminus K$ is path-connected as we can deduce from the figure and will prove in general in Proposition 4.1. Hence, the rays r_1 and r_2 are equivalent and define the same end. In particular, X has exactly one end and hence the end is nonplanar. In Proposition 3.10, we will show more generally that the number of singularities and the number of ends have to be the same if the metric completion is compact.

Altogether, the Chamanara surface is a Loch Ness monster which was stated previously without proof.

2.2 Exponential surface

Definition Consider the generalized polygon as in Figure 2.5 with the vertices $\{(n, \pm 2^{-|n|}) : n \in \mathbb{Z}\}$. It consists of infinitely many trapezoids where the nth trapezoid goes from $(n, \pm 2^{-|n|})$ to $(n+1, \pm 2^{-|n+1|})$ for $n \in \mathbb{Z}$. For each edge there exists a parallel edge of the same length, these pairs are glued to obtain the exponential surface $(X, \mathcal{A})$. Note that this gluing causes that pairs of trapezoids produce vertical cylinders.

Singularities We will now study the singularities of the exponential surface. Following the gluings, we find that the vertices are identified to at most two points (see Figure 2.6). As in the Chamanara surface, the distance of these two points in $\overline{X}$ would be 0, so it is only one point. Hence, the exponential surface has exactly one singularity σ. In particular, there exist arbitrarily short saddle connections from σ to itself. This means that there does not exist an $\epsilon > 0$ so that $B(\sigma, \epsilon) \setminus \{\sigma\}$ is isometric to a cyclic translation covering of a once-punctured disk, neither a finite covering nor an infinite one. Therefore, σ is a wild singularity. The two

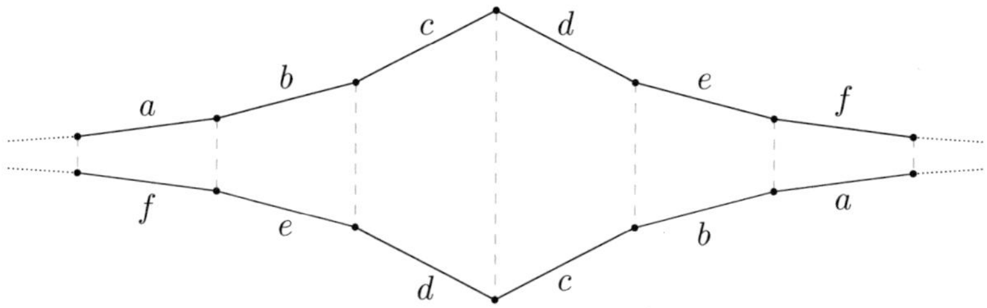

Figure 2.5: For the exponential surface, we identify parallel edges.

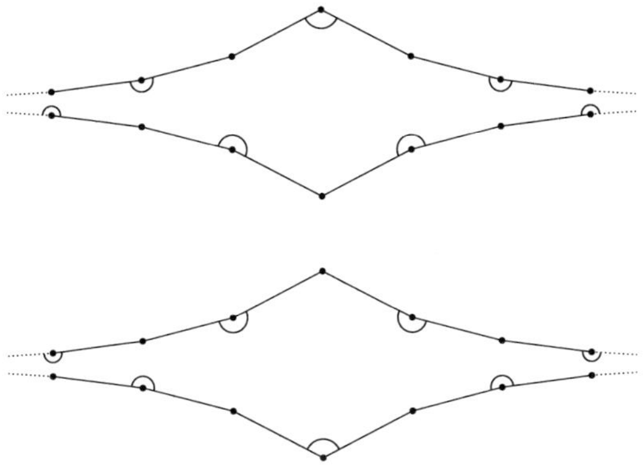

Figure 2.6: The arcs indicate which vertices of the generalized polygon are identified by the gluings in the exponential surface.

ways to go around σ give us two rotational components of σ which are both isometric to $\mathbb{R}$. There are no more rotational components as we will see implicitly from the next paragraph.

Spaces of linear approaches and rotational components For a linear approach $[\gamma]$ to σ, a neighborhood in $\mathcal{L}(\sigma)$ looks like this: let γ be a representative of $[\gamma]$ and $n \in \mathbb{Z}$ so that $\gamma((0, 2^{-|n|-2}))$ is contained in the trapezoid from $(n, \pm 2^{-|n|})$ to $(n+1, \pm 2^{-|n+1|})$. Suppose $\gamma((0, 2^{-|n|-2}))$ is not contained in the boundary of a trapezoid (even if so, the argument is similar). Then there exists an $r \in (0, 2^{-|n|-3})$ so that $B(\gamma(2^{-|n|-3}), r)$ is contained in the trapezoid which implies that $B(\gamma, 2^{-|n|-3}, r)$ is a neighborhood of $[\gamma]$ which only contains linear approaches that start in the same corner of the trapezoid. In fact, the directions of the linear approaches in $B(\gamma, 2^{-|n|-3}, r)$ only differ in a small angle from the direction of $[\gamma]$ which means they are close to $[\gamma]$ in the translation structure on $\overline{[\gamma]}$.

We have the following characterization of the topology on $\mathcal{L}(\sigma)$.

Proposition 2.3 (Topology of $\widetilde{\mathcal{L}}(\sigma)$)
Let $(X, \mathcal{A})$ be the exponential surface and σ the unique singularity. For both rotational components of σ, the subspace topology induced by the topology on $\mathcal{L}(\sigma)$ equals the topology induced by the one-dimensional translation structure on the rotational component defined in Definition 1.19.

For other translation surfaces, the statement of Proposition 2.3 is in general not true as we will point out in the discussion of the stack of boxes in Section 2.3.

The topology on the two-point space $\widetilde{\mathcal{L}}(\sigma)$ of rotational components is the discrete topology.

Cylinder decomposition There exists a vertical cylinder decomposition of the surface, with each cylinder consisting of two trapezoids. The nth cylinder has height 1 and circumference $3 \cdot 2^{-n}$ for $n \in \mathbb{N} = \{0, 1, \ldots\}$. Therefore, every inverse modulus is an integer multiple of $\frac{1}{3}$, i.e. the cylinder decomposition has commensurable inverse moduli. We can deduce from Proposition 1.38 that there exists a parabolic element $\left(\begin{smallmatrix} 1 & 0 \\ 3 & 1 \end{smallmatrix}\right)$ in the Veech group. It acts by twisting the nth cylinder 2^n times for $n \in \mathbb{N}$.

When defining cylinder decompositions as in Definition 1.36, there does not exist a horizontal cylinder decomposition on the exponential surface. However, it could be reasonable to define a cylinder decomposition as a collection of disjoint maximal cylinders so that the closure of the union of the cylinders covers the translation surface. But even if we would use this definition, for the horizontal cylinder decomposition of the exponential surface, the inverse moduli were not commensurable: the nth cylinder has height 2^{-n-1} and circumference $4n + 2$ for $n \in \mathbb{N}$, hence the inverse moduli are going to 0.

Veech group By using the vertical cylinder decomposition, we can show that the Veech group of the exponential surface is generated by a parabolic element and $-\mathrm{Id}$ which corresponds to the reflection through the origin.

Proposition 2.4 (Veech group of exponential surface)
For the exponential surface $(X, \mathcal{A})$, it holds

$$\mathrm{GL}^+(X, \mathcal{A}) = \left\langle \begin{pmatrix} 1 & 0 \\ 3 & 1 \end{pmatrix}, \begin{pmatrix} -1 & 0 \\ 0 & -1 \end{pmatrix} \right\rangle.$$

Proof. Note that every saddle connection of length less than 1 is vertical. This is because it cannot pass through a vertical cylinder (as it would need to have length greater than the height of the cylinder which is 1).

Now let φ be an orientation-preserving affine map, v a vertical geodesic of length 1, and λ the length of $\varphi(v)$. Then there exists a vertical saddle connection s which is shorter than $\frac{1}{\lambda}$. Therefore, $\varphi(s)$ is a saddle connection shorter than 1 and hence vertical. However, as the lengths of vertical saddle connections are bounded from above and every length arises at most twice, λ is 1 and φ^2 maps every vertical saddle connection to itself.

An element in $\mathrm{SL}(2,\mathbb{R})$ with vertical eigen direction and eigenvalue 1 is a parabolic element which twists every vertical cylinder at least one time. In this case, the derivative of φ is a power of $\left(\begin{smallmatrix}1&0\\3&1\end{smallmatrix}\right)$.

As the eigenvalue could also be -1, the derivative of φ is contained in

$$\left\langle \begin{pmatrix}1&0\\3&1\end{pmatrix}, \begin{pmatrix}-1&0\\0&-1\end{pmatrix} \right\rangle. \qquad \square$$

Topology of the surface We have an explicit set of nonseparating curves in the exponential surface, namely the core curves of the vertical cylinders, i.e. closed vertical curves that are contained in one cylinder. These curves form a nonseparating set of infinite cardinality. Hence, the exponential surface has infinite genus.

For the determination of the space of ends we use the same strategy as in the case of the Chamanara surface: Let $K \subseteq X$ be a compact set. Then there exist $n \in \mathbb{N}$ and $\epsilon > 0$ so that the set U which is defined as the union of the trapezoids from $(-n, \pm 2^{-n})$ to $(n, \pm 2^{-n})$ without $B(\sigma, \epsilon)$ contains K. Every proper ray has to stay in the path-connected set $X \setminus U$ (which contains $B(\sigma,\epsilon) \setminus \{\sigma\}$) after finite time, hence the exponential surface has exactly one end. Because of the infinite genus, the end is necessarily nonplanar. This shows that the exponential surface is also a Loch Ness monster.

Area and noncompactness The trapezoid from the vertex $(n, \pm 2^{-n})$ to the vertex $(n+1, \pm 2^{-n-1})$ for $n \in \mathbb{N}$ has height 1 and bases of length 2^{-n+1} and 2^{-n}, hence the area of the nth trapezoid can be calculated to be $\frac{1}{2}(2^{-n+1} + 2^{-n}) \cdot 1 = 3 \cdot 2^{-n-1}$. Therefore, the area of X is $2 \cdot 3 \cdot \sum_{n=0}^{\infty} \frac{1}{2^{n+1}} = 6$. However, it is not a finite translation surface as the metric completion is not compact. Indeed, we can choose an open strip in each maximal vertical cylinder, along the core curves of the cylinders. The collection of open strips can be extended to an open cover of X which does not have a finite subcover.

Modification with infinitely many discrete singularities We can modify the exponential surface so that it has infinitely many discrete singularities and the area is still finite: For every $n \in \mathbb{Z}$, consider a segment from $(n+\frac{1}{4}, 0)$ to $(n+\frac{3}{4}, 0)$ and a rectangle of width $\frac{1}{2}$ and height $\frac{1}{|n|}$. Glue the top of the rectangle to the lower part of the segment, the bottom of the rectangle to the upper part of the segment, and the left and the right edge of the rectangle together. This construction yields a cone angle singularity of multiplicity 3 for every $n \in \mathbb{Z}$. The distance of each two of these singularities is at least $\frac{1}{2}$ and the distance of each cone angle singularity to the wild singularity is at least $\frac{1}{4}$. Hence, the set of singularities is discrete.

2.3 Stack of boxes

History This family of examples was first described by Bowman in [Bow12]. He uses the family to produce concrete examples which show that different finiteness conditions on translation surfaces do not imply each other. We will use a modification of these examples later in the proof of Theorem 5.

Definition Choose two sequences $H = \{h_n\}$ and $W = \{w_n\}$ of positive real numbers so that H is bounded away from 0 and W is strictly monotonic decreasing and converges to 0. Now consider for each $n \in \mathbb{N}$ a rectangle R_n of height h_n and width w_n (called the nth box). We glue the boxes by identifying the right side of R_n with the left side of R_n. Furthermore, we identify the lower edge of R_{n+1} with the left part of the upper side of R_n. The lower side of R_0 is divided into segments of length $w_n - w_{n+1}$ and the right part of the upper side of R_n is glued to the corresponding segment on the lower side of R_0 (see Figure 2.7 for a sketch).

Singularities As all vertices of the rectangles and all points A_i are identified, the stack of boxes has exactly one singularity σ. This singularity has one rotational component which is isometric to $(0, \infty)$. We can specify the isometry by starting with a linear approach which has the left lower corner A_0 of R_0 as a base point and stays in R_0 for some time. The linear approach forms an angle $\theta \in (0, \frac{\pi}{2})$ with the horizontal

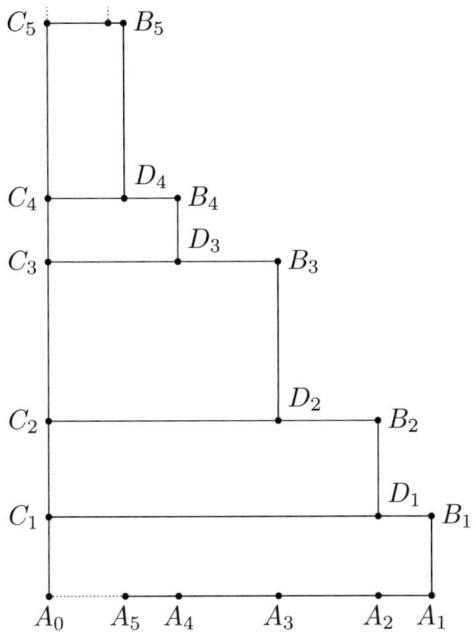

Figure 2.7: For the stack of boxes, we identify A_0C_1 with A_1B_1 and for each $n \geq 1$ we identify C_nC_{n+1} with D_nB_{n+1} and A_nA_{n+1} with B_nD_n.

direction and is sent to θ by the isometry. When rotating this linear approach counterclockwise around A_0, we obtain linear approaches with base points A_1, then $B_1, C_1, D_1, A_2, B_2, \ldots$ When we start the rotation clockwise we cannot rotate further than to an almost horizontal linear approach which corresponds to 0 not being contained in the image of the isometry. This discussion also shows that the singularity is wild.

Space of linear approaches As the only rotational component is isometric to $(0, \infty)$, we can identify each linear approach with a number $y \in (0, \infty)$. Let $[\gamma_y]$ be the corresponding linear approach and γ_y a representative. Then for each $y_0 \in (\pi, \infty)$, there exist $t > 0$ and $r > 0$ such that

$$B(\gamma_{y_0}, t, r) = \{[\gamma_y] : y \in (y_0 - \epsilon, y_0 + \epsilon)\}$$

for an $\epsilon := \epsilon(t, r) > 0$. Hence the subspace $\{[\gamma_y] : y \in (\pi, \infty)\} \subseteq \mathcal{L}(\sigma)$ carries the same topology as (π, ∞) with the Euclidean metric.

However, for a $y_0 \in (0, \pi]$ any neighborhood of $[\gamma_{y_0}]$ also contains linear approaches that have base point A_n for n large enough. In fact, for $t > 0$ and $r > 0$ small enough, we have

$$B(\gamma_{y_0}, t, r) = \{[\gamma_y] : y \in (y_0 - \epsilon, y_0 + \epsilon) \cup \bigcup_{n \geq N} (4\pi n + y_0 - \epsilon_n^+, 4\pi n + y_0 + \epsilon_n^-)\}$$

for an $N \geq 1$ and some $\epsilon := \epsilon(t, r)$, $\epsilon_n^+ := \epsilon_n^+(t, r)$, $\epsilon_n^- := \epsilon_n^-(t, r) > 0$ for all $n \geq N$. Hence, $\mathcal{L}(\sigma)$ does not carry the same topology as $(0, \infty)$ with the Euclidean metric.

We know from [BV13, Corollaries 2.9 and 2.11] that $\mathcal{L}(X)$ is T_2 (i.e. a Hausdorff space) and second-countable hence $\mathcal{L}(\sigma)$ is T_2 and second-countable. However, it is not T_3 as we show in the following lemma.

Lemma 2.5 ($\mathcal{L}(\sigma)$ is not T_3 and not metrizable). For the only singularity σ in the stack of boxes, $\mathcal{L}(\sigma)$ is not T_3. In particular, $\mathcal{L}(\sigma)$ is not metrizable.

Proof. For every $\epsilon > 0$ let ρ_ϵ be the embedding $\mathcal{L}^\epsilon(\sigma) \hookrightarrow \mathcal{L}(\sigma)$ from the direct system. Now fix $\epsilon' > 0$ and define the set

$$F = \mathcal{L}(\sigma) \setminus \Big(\bigcup_{\epsilon > \epsilon'} \rho_\epsilon (\mathcal{L}^\epsilon(\sigma)) \Big).$$

This is a closed set as $\bigcup_{\epsilon > \epsilon'} \rho_\epsilon(\mathcal{L}^\epsilon(\sigma))$ is open in $\mathcal{L}(\sigma)$ (see [BV13, Corollary 2.5] for a proof).

For ϵ' small enough, the horizontal linear approach $[\gamma_\pi]$ with base point A_1 is not contained in F. However, F contains $[\gamma_{(4n+1)\pi}]$ for n large enough and hence an open neighborhood of F contains $[\gamma_{(4n+1)\pi - \epsilon_n}]$ for n large enough and $\epsilon_n > 0$ small enough. Furthermore, every open neighborhood of $[\gamma_\pi]$ contains $[\gamma_{(4n+1)\pi - \epsilon_n}]$ for n large enough and $\epsilon_n > 0$ small enough as we described before. All in all, every open neighborhood of F has nonempty intersection with every open neighborhood of $[\gamma_\pi]$. Hence, $\mathcal{L}(\sigma)$ is not T_3.

As every metric space fulfills T_3, this shows in particular that $\mathcal{L}(\sigma)$ is not metrizable. □

2.4 Slit constructions

If we are interested in translation surfaces with given properties then the slit construction described in this section is very helpful. We choose a translation surface with desirable properties, cut it along geodesic segments and reglue the segments in a different way. By performing this operation carefully, we are able to control some of the properties of the resulting translation surface, for example the area does not change.

This was partly done with Lucien Clavier and Chenxi Wu and previously published in [CRW14]. The method of slit constructions was also described before as *regluing of marks* in [PSV11, Definitions 3.4 and 3.6].

General construction Let $(X_1, \mathcal{A}_1)$ and $(X_2, \mathcal{A}_2)$ be translation surfaces that are not necessarily different. A *slit* m on $(X_1, \mathcal{A}_1)$ is a geodesic segment in X_1 with an orientation. When we consider the metric completion of $X_1 \setminus m$ equipped with the path length metric, we obtain two copies m' and m'' of m, with end points identified. If m_1 is a slit on $(X_1, \mathcal{A}_1)$ and m_2 is a slit on $(X_2, \mathcal{A}_2)$ with the same holonomy vector then we can "glue the slits together". For instance, we can identify m_1' with m_2'' or m_1'' with m_2' by a translation and we describe the first identification visually as "gluing the upper part of m_1 to the lower part of m_2". If we glue both, m_1' with m_2'' and m_1'' with m_2', we say that we "glue m_1 and m_2 (to each other)".

Another possibility is to consider one slit m and divide m' and m'' into several segments which we reglue in a different order as the original one.

In both cases, the end points of the slits or the end points of the segments on the slit are no longer regular points, so we exclude them from the resulting surface.

Geometric series decoration The geometric series decoration can be performed on every translation surface. However, to keep notation simple, we consider the Euclidean plane $\mathbb{R}^2$ with a slit from $(0,0)$ to $(1,0)$ (cf. [BV13, Example 1.12]). We cut the upper and the lower part of the slit into halves and glue the top left segment to the bottom right segment. Again we cut the remaining parts into halves and glue the left half on the top to the right half on the bottom, and so on. We then have for every

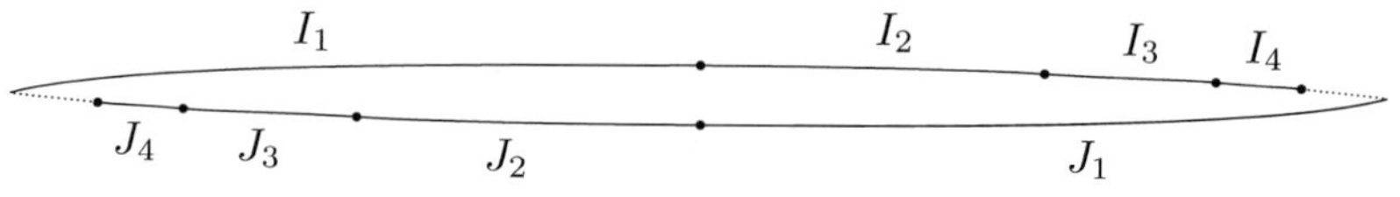

Figure 2.8: For the geometric series decoration on $\mathbb{R}^2$, segments I_n and J_n are glued (to illustrate the slit in $\mathbb{R}^2$, it is stretched out a bit).

$n \geq 1$ a segment I_n on the upper part of the slit and a segment J_n on the lower part of the slit, both of length 2^{-n} (see Figure 2.8).

The resulting surface has exactly one singularity σ. The singularity has two rotational components and both are isometric to $(0, \infty)$. Now let $[\gamma_1]$ be the vertical linear approach with base point the far right of the slit and going upward. Its corresponding rotational component shall be $\overline{[\gamma_1]}$ and the other rotational component is called $\overline{[\gamma_2]}$. Then every vertical linear approach with base point the right end point of I_{2n} for $n \geq 1$ and going upward is contained in $\overline{[\gamma_2]}$. Thus there exists a sequence of vertical linear approaches contained in $\overline{[\gamma_2]}$ that converges to $[\gamma_1]$ in $\mathcal{L}(\sigma)$. We can conclude that every open set in $\widetilde{\mathcal{L}}(\sigma)$ containing $\overline{[\gamma_1]}$ also contains $\overline{[\gamma_2]}$, and by symmetry we see that $\widetilde{\mathcal{L}}(\sigma)$ carries the trivial topology $\left\{\emptyset, \{\overline{[\gamma_1]}, \overline{[\gamma_2]}\}\right\}$.

Geometric series decoration (modified) We can modify the previous example by arranging the segments I_n and J_n in another order: we first assign a segment I_1 of length 2^{-1} to the left of the upper part of the slit and then successively assign segments I_n of length 2^{-n} for every $n \geq 2$ on its right. For the lower part, we assign a segment J_2 of length 2^{-2} at the left, a segment J_1 of length 2^{-1} on its right, then J_4 of length 2^{-4}, J_3 of length 2^{-3}, and so on. Then we glue the segments of the same length (see Figure 2.9).

Again, we have one singularity σ and two rotational components. However, one rotational component has infinite length and is isometric to $(-\infty, \infty)$ and one has finite length and is isometric to $(-\pi, \pi)$.

With a similar argument as in the original geometric series decoration, we see that the rotational component $\overline{[\gamma_{inf}]}$ of infinite length contains linear approaches which converge to a linear approach contained in the rotational component $\overline{[\gamma_{fin}]}$ of finite length. However, there is no sequence

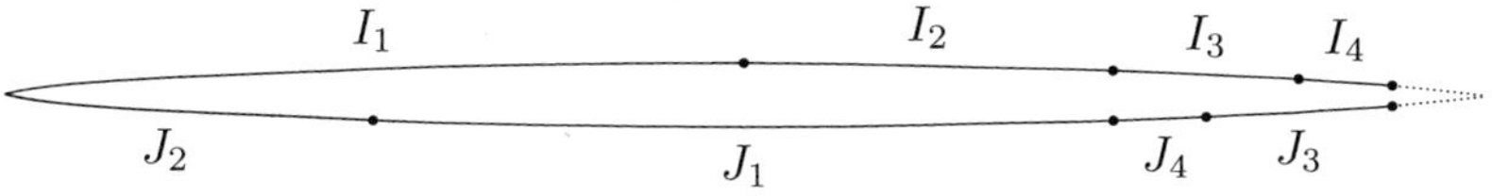

Figure 2.9: For the modified geometric series decoration, the segments on the upper and the lower part of the slit are going in the same direction.

in $\overline{[\gamma_{fin}]}$ converging to a linear approach contained in $\overline{[\gamma_{inf}]}$. Therefore, the topology of $\widetilde{\mathcal{L}}(\sigma)$ is $\{\emptyset, \{\overline{[\gamma_{inf}]}\}, \{\overline{[\gamma_{\text{inf}}]}, \overline{[\gamma_{fin}]}\}\}$.

Harmonic series decoration Consider a Euclidean plane $\mathbb{R}^2$ with an infinite slit from $(0,0)$ in horizontal direction. Divide the upper part of the slit into the following segments. For $n \geq 1$, let I_n be the segment from $(\sum_{k=1}^{n-1} \frac{1}{k}, 0)$ to $(\sum_{k=1}^{n} \frac{1}{k}, 0)$, i.e. the segment I_n has length $\frac{1}{n}$. Furthermore, divide the lower part of the slit into the following segments. If n is even, let J_n be the segment from $(\sum_{k=1}^{n-2} \frac{1}{k}, 0)$ to $(\sum_{k=1}^{n-2} \frac{1}{k} + \frac{1}{n}, 0)$. If n is odd, let J_n be the segment from $(\sum_{k=1}^{n-1} \frac{1}{k} + \frac{1}{n+1}, 0)$ to $(\sum_{k=1}^{n+1} \frac{1}{k}, 0)$. This means that the length of J_n is also $\frac{1}{n}$ but the order is different from the order of the segments I_n (cf. the modified geometric series decoration). Then we identify I_n with J_n for every $n \geq 1$ (see Figure 2.10).

In this case, all the end points of the segments are identified via the gluings. Hence, the resulting translation surface has one singularity σ. It has exactly one rotational component which is isometric to $\mathbb{R}$. However, there does not exist an $\epsilon > 0$ so that $B(\sigma, \epsilon)$ is a locally flat neighborhood of σ. In the presentation as a Euclidean plane with a slit, one can observe that infinitely many of the glued segments are contained in any $B(\sigma, \epsilon)$. This means that for every $\epsilon > 0$ there exists a saddle connection of length

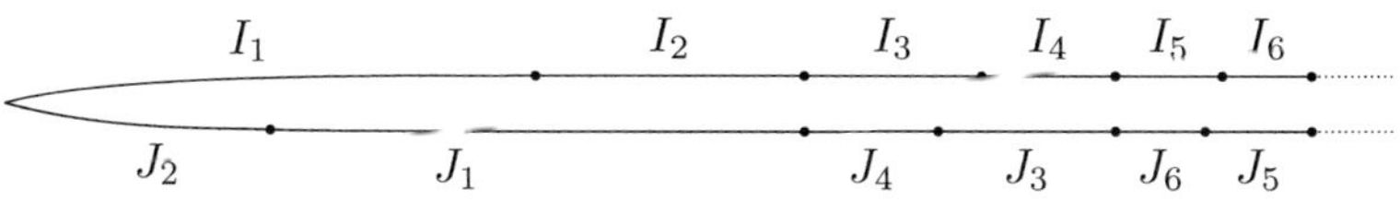

Figure 2.10: For the harmonic series construction on $\mathbb{R}^2$, the segments I_i and J_i are glued (to illustrate the slit in $\mathbb{R}^2$, it is stretched out a bit).

less than ϵ starting and ending at σ. In particular, there does not exist a punctured neighborhood of the singularity which is a translation covering of $B(0, \epsilon) \setminus \{0\} \subseteq \mathbb{R}^2$. Hence, the singularity is wild.

For each linear approach, there exists a neighborhood in the one-dimensional translation structure on the rotational component which is also a neighborhood of the linear approach in $\mathcal{L}(\sigma)$. Therefore, the topology on $\mathcal{L}(\sigma)$ is the same as the Euclidean topology on $\mathbb{R}$. Obviously, the topology of $\widetilde{\mathcal{L}}(\sigma)$ is already determined as it is the one-point space.

Note that the harmonic series decoration can not be performed on a torus as the slit would be dense in the torus and the dense occurrence of the singularity would make it impossible to define charts on the torus without the singularity.

Star decoration (without cylinders) Consider a translation surface $(X, \mathcal{A})$ and a regular point $x \in X$. We cut open every slit starting at x, of length 2^{-n}, and in the direction $\frac{m\pi}{2^n}$ such that $n \geq 1$ and $m \in \mathbb{N}$ is odd when $n > 1$ (see Figure 2.11). The directions $\frac{m\pi}{2^n}$ are called *dyadic* whereas all other directions are called nondyadic. We will describe two ways to glue the branches of this star (the second will involve additional cylinders) which result in translation surfaces with non-homeomorphic spaces of rotational components.

The first kind of gluing is the following. For each branch, glue the right side to the left side of the antipodal branch and vice versa. Then all tips of the branches are identified with the center and there is exactly one singularity, which is wild. For every branch, we have a rotational component which is isometric to $[0, 2\pi]$. Also, for every nondyadic direction, we have a linear approach with base point the center of the star. It is contained in a rotational component that consists only of this point.

The difficulty with this gluing is of the same kind as in the Chamanara example: there might be additional linear approaches that are not easy to see in the figure as they are intersecting infinitely many branches. It is possible to find such a linear approach by starting with any regular point, and considering geodesics passing through that point. We can then inductively find nested open sets U_n of directions for which the geodesics intersect n smaller and smaller branches. By construction the intersection of these nested open intervals is a singleton, consisting of a direction for which the geodesic reaches the center of the star in finite time (i.e. defines

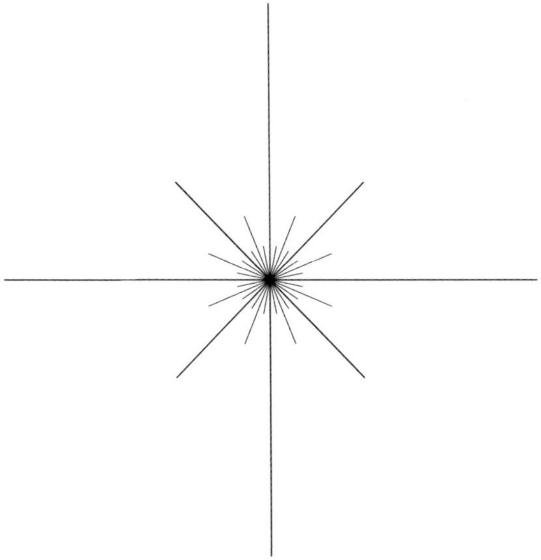

Figure 2.11: Branches of the star in the star decoration.

a linear approach) while passing through infinitely many branches. In Figure 2.12, a representative of such a linear approach is indicated.

Because of the additional linear approaches, the topology of $\mathcal{L}(\sigma)$ is hard to describe, so we modify the gluings of the branches.

Star decoration (with cylinders) Consider a translation surface $(X, \mathcal{A})$ with slits as before. For every branch of the star, consider a rectangle of the same height as the branch and of width 1 and rotate it so that the former vertical sides are parallel to the branch. Then glue the former left side of the rectangle to the right side of the branch, the former right side of the rectangle to the left side of the branch, and identify the former top and bottom of the rectangle.

Note that the resulting translation surface does not have a compact metric completion, even if the original translation surface $(X, \mathcal{A})$ has a compact metric completion. If we are interested in a translation surface with compact metric completion then we use rectangles as in Figure 2.13

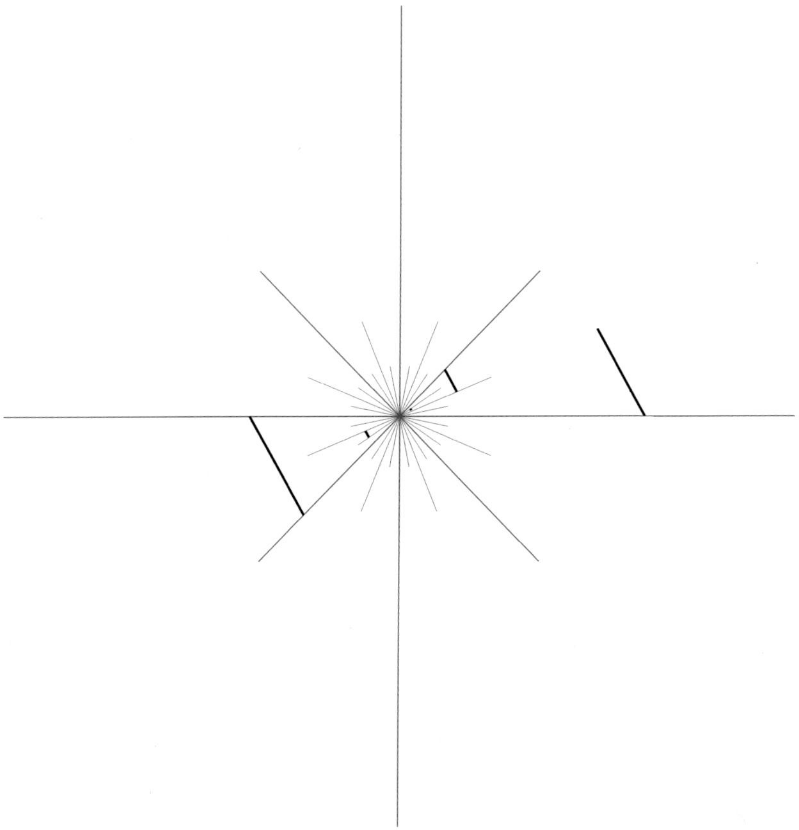

Figure 2.12: In the star decoration without additional cylinders, the geodesic curve starts in the center of the star and intersects infinitely many branches.

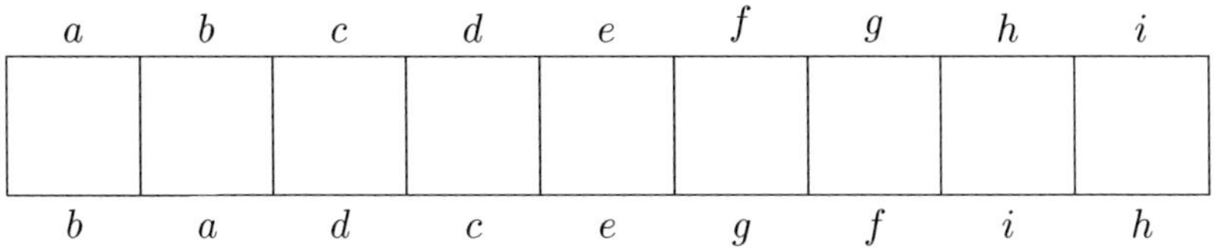

Figure 2.13: A rectangle consisting of $2^n + 1 + 2^n$ squares for $n = 2$: segments with the same letters are glued, the vertical edges are glued to the corresponding branch of the star.

instead of the standard rectangles. This type of rectangles consists of $2^n + 1 + 2^n$ squares of side length 2^{-n}, with the tops and bottoms of adjacent squares glued crosswise, except for the middle square where top and bottom are identified. Then every open cover of the metric completion includes a neighborhood of the singularity σ and this neighborhood contains $B(\sigma, \epsilon)$ for an $\epsilon > 0$. Hence, every additional rectangle with height less than ϵ is contained in the neighborhood of σ and only finitely many additional rectangles of greater height remain to be covered by a finite subcover.

As in the previous case, we have exactly one singularity σ as all tips of the branches are identified with the center as well as the vertices of the squares in each rectangle. By the gluing-in of the additional rectangles, we excluded linear approaches that intersect infinitely many branches, hence there are only four types of linear approaches with the following types of representatives.

(nd) Geodesic segments γ starting from the center of the star with a nondyadic direction, and such that $\gamma(t)$ is not in the interior of a rectangle for small t,

(ct) geodesic segments γ starting from the center of the star with a dyadic direction, and such that $\gamma(t)$ is not in the interior of a rectangle for small t,

(tp) geodesic segments γ starting from the tip of a branch of the star, and such that $\gamma(t)$ is not in the interior of a rectangle for small t, (rt) geodesic segments γ starting at a vertex of a square and such that $\gamma(t)$ stays in the interior of the corresponding rectangle for small t.

Now there exist two types of rotational components (see Figure 2.14):

(N) singletons consisting of a linear approach of type (nd), and

(D) rotational components of finite length and isometric to the interval $[0,(2^{n+1}+2)\cdot 2\pi]$ via a map ϕ such that $\phi(0)$ and $\phi((2^{n+1}+2)\cdot 2\pi)$ are of type (ct) and have the same (dyadic) direction, $\phi(x)$ is of type (tp) for $x \in [(2^{n+1}+1)\cdot\pi,(2^{n+1}+3)\cdot\pi]$ and of type (rt) otherwise.

We will now study the topology of $\widetilde{\mathcal{L}}(\sigma)$. For this, let

$$P\colon \widetilde{\mathcal{L}}(\sigma) \to S^1 = \mathbb{R}/2\pi\mathbb{Z}$$

be the map that sends a rotational component to the direction of its element(s) of type (nd) or (ct). By definition, P is bijective. We will show that P is a homeomorphism in two steps by comparing the canonical topology of S^1 and the pushforward topology of $\widetilde{\mathcal{L}}(\sigma)$ on $P(\widetilde{\mathcal{L}}(\sigma))$. The first lemma indicates that the topology of $P(\widetilde{\mathcal{L}}(\sigma))$ is finer than the topology of S^1 and the second lemma indicates that the topology of $P(\widetilde{\mathcal{L}}(\sigma))$ is coarser than the topology of S^1.

Lemma 2.6 (Topology of $P(\widetilde{\mathcal{L}}(\sigma))$ is finer than topology of S^1). Let $(X,\mathcal{A})$ be a translation surface with a star decoration and σ the corresponding singularity. Furthermore, let $\overline{[\gamma]}$ be a rotational component in $\widetilde{\mathcal{L}}(\sigma)$ and $\epsilon \in S^1$. Then there exists an open neighborhood $U \subseteq \widetilde{\mathcal{L}}(\sigma)$ of $\overline{[\gamma]}$ such that $P(U) \subseteq (P(\overline{[\gamma]})-\epsilon, P(\overline{[\gamma]})+\epsilon)$.

Proof. As a preliminary observation note that linear approaches of type (ct) can not have representatives which are longer than the corresponding branch, in particular the length has to be less than $\frac{1}{2}$. Furthermore, the representative of a linear approach has to be longer than 1 to pass through a rectangle since it has to pass through the 2^n pairs of squares which each takes at least time 2^{-n}.

Now let $\overline{[\gamma]}$ be a rotational component and $\epsilon \in S^1$. We show in the following that we can find an open neighbourhood U of $\overline{[\gamma]}$ such that $P(U) \subseteq (P(\overline{[\gamma]})-\epsilon, P(\overline{[\gamma]})+\epsilon)$. This is done by studying the possible neighborhoods of each linear approach $[\gamma]$ which is contained in the rotational component $\overline{[\gamma]}$ in a case-by-case analysis.

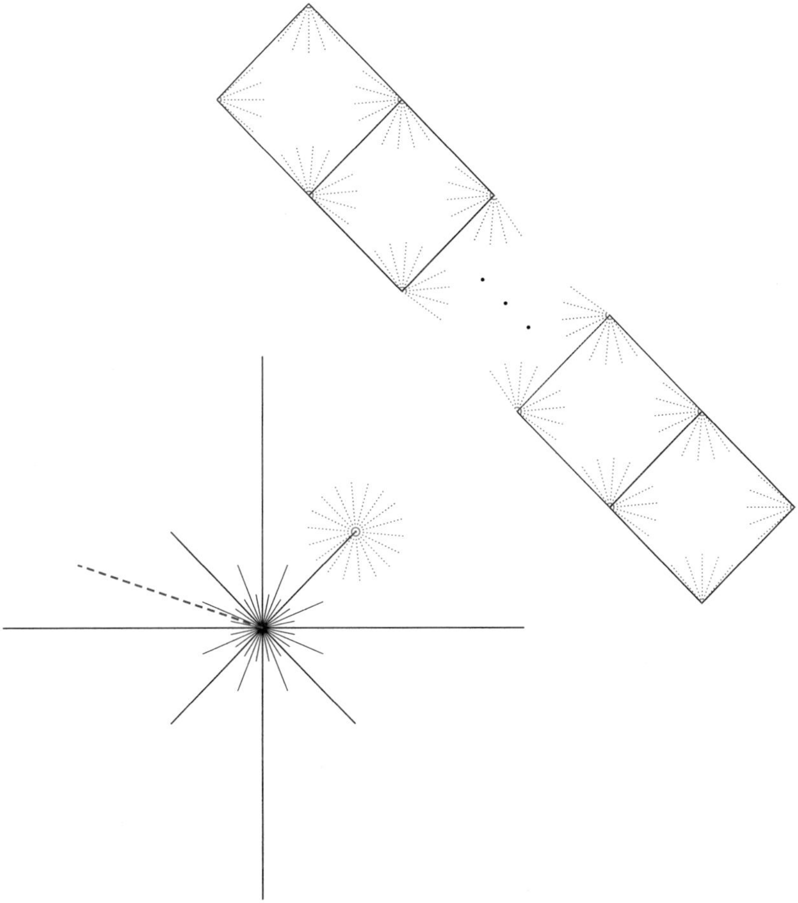

Figure 2.14: Two types of rotational components in the star decoration with additional cylinders: the dashed segment indicates a linear approach contained in a rotational component of type (N), the dotted segments indicate linear approaches contained in a rotational component of type (D).

(nd) Suppose that $[\gamma]$ is a linear approach with a representative γ of type (nd). Choose $t = 1$ and $r > 0$ small enough so that $B(\gamma, t, r)$ does not contain linear approaches of type (ct) and (rt).

Now let γ' be a representative of a linear approach of type (nd) which is contained in $B(\gamma, t, r)$. Then by the law of cosines we have that "the angle between γ' and γ" is

$$\left|P(\overline{[\gamma]}) - P(\overline{[\gamma']})\right| \leq \arccos\left(\frac{r^2 - t^2 - t^2}{-2 \cdot t \cdot t}\right) = \arccos\left(1 - \frac{r^2}{2t^2}\right).$$

This means that by shrinking r we can bound the distance of $P(\overline{[\gamma]})$ and $P(\overline{[\gamma']})$ in S^1 by any given value greater than 0, in particular by the given ϵ.

We also want to bound this distance for rotational components of type (D). By the choice of t and r, we only have to consider linear approaches of type (tp). Choose two branches of the star whose directions are contained in $(P(\overline{[\gamma]}) - \epsilon, P(\overline{[\gamma]}))$ and $(P(\overline{[\gamma]}), P(\overline{[\gamma]}) + \epsilon)$, respectively. Furthermore, choose $t' > 0$ small enough so that no branch with direction not in $(P(\overline{[\gamma]}) - \epsilon, P(\overline{[\gamma]}) + \epsilon)$ intersects one of the geodesic rays which start in $\gamma(t')$ and pass through the tip of one of the chosen branches (see Figure 2.15). Moreover, let $r' > 0$ be so small that $B(\gamma(t'), r')$ does not intersect any branch. Then every linear approach $[\gamma']$ of type (tp) with $P(\overline{[\gamma]}) - P(\overline{[\gamma']}) \geq \epsilon$ that is contained in $B(\gamma, t', r')$ has a representative γ' which passes through at least one cylinder (namely one of the cylinders glued in at the chosen branches) before entering $B(\gamma(t'), r')$. However, t' can be chosen to be less than 1 so $\gamma'((0, t'))$ can not pass through a cylinder which implies that no such γ' exists.

In summary, this means that $V := B(\gamma, t, r) \cap B(\gamma, t', r')$ is a neighborhood of $[\gamma]$ which contains only linear approaches of type (nd) and (tp). Let U be the projection of V to $\widetilde{\mathcal{L}}(\sigma)$. Then we have shown in the discussion above (and with an anticipation of the cases (tp) and (rt) below) that $P(\overline{[\gamma']})$ has distance at most ϵ from $P(\overline{[\gamma]})$ for every $\overline{[\gamma']} \in U$.

This proves the statement for rotational components $\overline{[\gamma]}$ of type (N).

(ct) Suppose that $[\gamma]$ is a linear approach with a representative γ of type (ct). Let $t \in (0, \frac{1}{2})$ and $r > 0$ small enough so that $B(\gamma(t), r)$

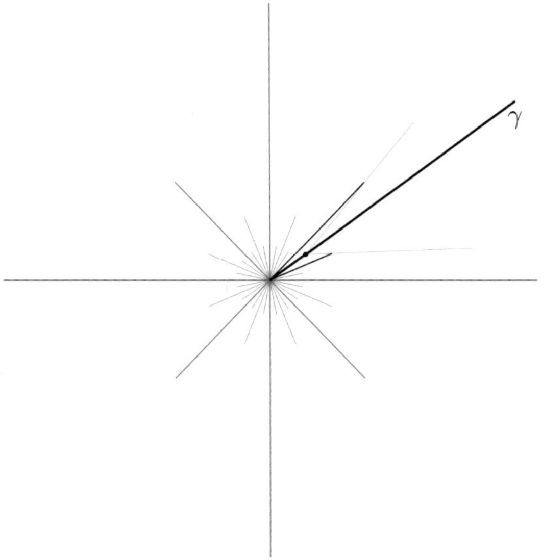

Figure 2.15: The constellation in the case (nd) for the choice of t'.

does not intersect any other branches. This implies, that no linear approach of type (ct) except $[\gamma]$ is contained in $B(\gamma, t, r)$. Again, by the law of cosines, the distance of the images of the corresponding rotational components for all linear approaches in $B(\gamma, t, r)$ of type (nd) is bounded in terms of r. Furthermore, every linear approach of type (rt) contained in $B(\gamma, t, r)$ is contained in the same rotational component as $[\gamma]$. Also, for linear approaches of type (tp), the same argument as in the previous case works.

(tp) Suppose that $[\gamma]$ is a linear approach with a representative γ of type (tp), with base point the tip of a branch of length 2^{-n}. Choose $t > 0$ and $r > 0$ small enough so that $2t + r < 2^{-n}$ and so that $B(\gamma(t), r)$ does not intersect any branches. Then $B(\gamma, t, r)$ can not contain linear approaches of type (nd) or (ct) by the first condition and can not contain linear approaches of type (rt) by the first and the second condition.

By choosing $t' < \frac{t}{2}$ and $r' > 0$ small enough, we see that the intersection $B(\gamma, t, r) \cap B(\gamma, t', r')$ only contains linear approaches of type (tp) and so that all of these linear approaches are contained in the same rotational component as $[\gamma]$.

(rt) Suppose that $[\gamma]$ is a linear approach of type (rt). By combining arguments from the previous three cases we see that there exists an open neighborhood of $[\gamma]$ that only contains linear approaches of type (rt) which are contained in the same rotational component as $[\gamma]$: For $t > 0$ and $r > 0$ small enough so that $B(\gamma(t), r)$ does not intersect any edges of the rectangle, the neighborhood $B(\gamma, t, r)$ can not contain linear approaches of type (nd) or (ct). By choosing $t' < \frac{t}{2}$ and $r' > 0$ small enough, we can again make sure that $B(\gamma, t, r) \cap B(\gamma, t', r')$ does not contain linear approaches of type (tp) and no linear approaches of type (rt) that are not contained in the same rotational component as $[\gamma]$.

In fact, the discussion of the cases (ct), (tp), and (rt) shows that for a rotational component $\overline{[\gamma]}$ of type (D), every linear approach contained in $\overline{[\gamma]}$ has a neighborhood N so that for each linear approach $[\gamma'] \in N$ it holds $|P(\overline{[\gamma']}) - P(\overline{[\gamma]})| < \epsilon$. Let V be the union of these neighborhoods and U the projection of V to $\widetilde{\mathcal{L}}(\sigma)$. Because of the discussion above, we know that the preimage of U differs from V by some additional linear approaches of type (tp) and (rt) that are contained in rotational components in U. Therefore, the preimage of U is open in $\mathcal{L}(\sigma)$ and hence U is open in $\widetilde{\mathcal{L}}(\sigma)$. Furthermore, we have $P(U) \subseteq (P(\overline{[\gamma]}) - \epsilon, P(\overline{[\gamma]}) + \epsilon)$ which was to be shown. □

We now exhibit a variant of a converse subset relation for the topology.

Lemma 2.7 (Topology of $P(\widetilde{\mathcal{L}}(\sigma))$ is coarser than topology of S^1). Let $(X, \mathcal{A})$ be a translation surface with a star decoration and σ the corresponding singularity. Furthermore, let $\overline{[\gamma]}$ be a rotational component in $\widetilde{\mathcal{L}}(\sigma)$ and $U \subseteq \widetilde{\mathcal{L}}(\sigma)$ an open neighborhood of $\overline{[\gamma]}$. Then there exists an $\epsilon \in S^1$ such that $(P(\overline{[\gamma]}) - \epsilon, P(\overline{[\gamma]}) + \epsilon) \subseteq P(U)$.

Proof. Let V be the preimage of U in $\mathcal{L}(\sigma)$. Again, we prove the statement by a case-by-case analysis.

(N) Let $\overline{[\gamma]}$ be a rotational component of type (N), i.e. there exists a linear approach $[\gamma]$ with a representative γ of type (nd). By Lemma 1.22 there exist $t > 0$ and $r > 0$ such that $B(\gamma, t, r)$ is contained in V. We can even choose $t > 1$ in this situation. By the same calculation as in the proof of the previous lemma, for a linear approach $[\gamma']$ of type (nd) with $\left|P(\overline{[\gamma]}) - P(\overline{[\gamma']})\right| < \arccos(1 - \frac{r^2}{2t^2})$ we have $[\gamma'] \in B(\gamma, t, r)$. Furthermore, the argument by the law of cosines again yields an $\epsilon \in (0, \arccos(1 - \frac{r^2}{2t^2}))$ so that $\gamma'(t) \in B(\gamma(t), r)$ for every representative γ' of type (tp) with direction equal to the direction $P([\gamma'])$ of the branch whose tip is its base point and $P([\gamma']) \in (P(\overline{[\gamma]}) - \epsilon, P(\overline{[\gamma]}) + \epsilon)$. As there exists a linear approach of type (nd) for each nondyadic direction and a linear approach of type (tp) for each dyadic direction in $(P(\overline{[\gamma]}) - \epsilon, P(\overline{[\gamma]}) + \epsilon)$, we have

$$\left(P(\overline{[\gamma]}) - \epsilon, P(\overline{[\gamma]}) + \epsilon\right) \subseteq P\left(\left\{\overline{[\gamma']} : [\gamma'] \in B(\gamma, t, r)\right\}\right) \subseteq P(U).$$

(D) Let $\overline{[\gamma]}$ be a rotational component of type (D) and let $[\gamma_1]$, $[\gamma_2]$ be the linear approaches in $\overline{[\gamma]}$ with representatives γ_1, γ_2 of type (ct). Again, there exist $t > 0$ and $r > 0$ such that $B(\gamma_1, t, r) \cup B(\gamma_2, t, r)$ is contained in V. Now we can conclude by the same arguments as in the case (N) that there exists an $\epsilon > 0$ as desired. □

The two preceding lemmas lead us to the following proposition.

Proposition 2.8 (Topology of $\widetilde{\mathcal{L}}(\sigma)$)
Let $(X, \mathcal{A})$ be a translation surface with a star decoration and σ the corresponding singularity. Then the space $\widetilde{\mathcal{L}}(\sigma)$ of rotational components is homeomorphic to S^1.

Proof. The two preceding lemmas show that the preimages of the elements of the canonical subbasis for S^1 under P form a subbasis for $\widetilde{\mathcal{L}}(\sigma)$. □

3 Ends of translation surfaces

At least for cone angle singularities, we know that the singularities can topologically be seen as punctures in the metric completion. On the other hand, punctures of surfaces are ends of surfaces as we noted in Section 1.3. So it is natural to ask whether the set of singularities (equipped with the subspace topology from the metric completion) and the space of ends are related, for example whether the two spaces are in one-to-one correspondence. We will study possible relations in this chapter.

3.1 Different types of ends

Recall from Definition 1.30 that a *planar end* of a surface X is an equivalence class of proper rays so that there exists a compact set K such that the corresponding path-connected component of $X \setminus K$ has genus 0. In the case of a translation surface, we define two more types of ends in a similar way.

Definition 3.1 (Regular and singular ends and ends of (in)finite area)
Let $(X, \mathcal{A})$ be a translation surface and $\overline{X}$ the metric completion of X.

(i) An end e of X is said to be a *regular end of* $(X, \mathcal{A})$ if there exist a proper ray r with $\text{end}(r) = e$, a compact set $K \subseteq X$, and an $n \in \mathbb{N}$ so that the path-connected component C of $X \setminus K$, in which $r([n, \infty))$ is contained, does not contain a singularity, i.e. $\overline{C} \subseteq X$.

Otherwise, the end e is said to be a *singular end of* $(X, \mathcal{A})$.

(ii) An end e of X is said to be an end of $(X, \mathcal{A})$ *of finite area* if there exist a proper ray r with $\text{end}(r) = e$, a compact set $K \subseteq X$, and an $n \in \mathbb{N}$ so that the path-connected component of $X \setminus K$, in which $r([n, \infty))$ is contained, has finite area.

Otherwise, the end e is said to be an end of $(X, \mathcal{A})$ *of infinite area*.

It is noticeable that there are two different types of singular ends: a singular end could be an end so that the metric completion of the corresponding path-connected components of $X \setminus K$ for a compact set K contains infinitely many singularities. Then again, it could be an end that corresponds to the puncture that corresponds to a singularity. We will formalize the description of the latter situation.

Let $(X, \mathcal{A})$ be a translation surface and σ a singularity. This means that there exists a Cauchy sequence in X which converges to σ in $\overline{X}$. Now let $(x_n)_{n\in\mathbb{N}}$ be such a Cauchy sequence. We can define a proper ray $r_{(x_n)}\colon [0,\infty) \to X$ in the following way. As X is path-connected, for each $n \in \mathbb{N}$ there exists a curve $r_n\colon [0,1] \to X$ which connects x_n to x_{n+1}. Define $r_{(x_n)}\colon [0,\infty) \to X$ as $r_{(x_n)}(t) = r_{\lfloor t \rfloor}(t - \lfloor t \rfloor)$. This is a proper ray as every compact set in X keeps a distinct distance to the singularity σ and hence the ray $r_{(x_n)}$ leaves every compact set. This leads to the following definition.

Definition 3.2 (End of a singularity)
Let $(X, \mathcal{A})$ be a translation surface, σ a singularity, $(x_n)_{n\in\mathbb{N}} \subseteq X$ a Cauchy sequence converging to σ, and $r_{(x_n)}\colon [0,\infty) \to X$ a proper ray as defined before.

Then the *end of* σ is defined to be $\mathrm{end}(\sigma) := \mathrm{end}(r_{(x_n)})$.

We have to make certain that the end of a singularity is well-defined. Indeed, let $(y_n)_{n\in\mathbb{N}}$ be another Cauchy sequence in X which converges to σ. This means that for every $\epsilon > 0$ there exists an $N \in \mathbb{N}$ so that $d(x_n, y_n) < \epsilon$ for each $n \geq N$. Let $r_{(y_n)}$ be the corresponding proper ray as defined above. Now suppose that K is a compact set in X and C is the path-connected component of $X \setminus K$ belonging to $\mathrm{end}(r_{(x_n)})$. Then $d := d(K, \sigma) > 0$ and $B(\sigma, d) \cap X \subseteq C$. Therefore, we have an $n_0 \in \mathbb{N}$ with $r_{(y_n)}([n_0, \infty)) \subseteq C$ which leads to $\mathrm{end}(r_{(x_n)}) = \mathrm{end}(r_{(y_n)})$.

Definition 3.3 (End induced by a singularity)
Let $(X, \mathcal{A})$ be a translation surface and e an end of X. We say that the end e is *induced by a singularity* if there exists a singularity σ such that $\mathrm{end}(\sigma) = e$.

The number of singularities that induce a given singular end can be 0, any finite number, or even infinite as we will see in the following example.

Example 3.4 (Number of singularities inducing an end).

(i) Consider a Euclidean plane with slits s_n from $(2n, 0)$ to $(2n+1, 0)$ and slits t_n from $(-2n-1, 0)$ to $(-2n, 0)$ for $n \geq 1$. When gluing s_n and t_n for every $n \geq 1$, we obtain a translation surface $(X, \mathcal{A})$ with infinitely many singularities (all of them cone angle singularities of multiplicity 2), infinite genus and infinitely many ends. The end which is defined by the ray $r\colon [0, \infty) \to X,\ x \mapsto (0, x)$ is a singular end. However, it is not induced by any of the singularities.

(ii) Recall the infinite staircase $(X, \mathcal{A})$ from Example 1.8 and consider a proper geodesic ray $r\colon [0, \infty) \to X$. Every compact set $K \subseteq X$ has to avoid a small neighborhood of the singularity and can only intersect finitely many of the squares. The "upper" and the "lower" part of the infinite staircase are path-connected via the avoided neighborhood of the singularity, hence $X \setminus K$ consists of one unbounded path-connected component and possibly some bounded path-connected components. The ray r is contained in the unbounded path-connected component after finite time as it has to leave every finite union of squares except of possibly a neighborhood of the singularity. This means that the space of ends consists exactly of $\operatorname{end}(r)$. In particular, the ends of all four infinite angle singularities are equal to $\operatorname{end}(r)$, so $\operatorname{end}(r)$ is induced by four singularities.

(iii) Let $n \geq 1$ and consider a Euclidean plane with n geodesic rays that do not intersect. Now we treat every geodesic ray as a slit, cut it into segments, and glue the segments as for the harmonic series decoration from Section 2.4. By this construction, we obtain a translation surface $(X, \mathcal{A})$ with n singularities which are all wild singularities. For a compact subset of X, the complement consists of one unbounded path-connected component and possibly some bounded path-connected components. By the same argument as in the previous example, $(X, \mathcal{A})$ has exactly one end and this end is induced by all n singularities.

The same construction also works when we replace the number n by "infinite". In this case, we have to make sure that the infinitely many singularities do not accumulate, for example by choosing horizontal rays with starting points equispaced on the vertical axis. Then we obtain a translation surface with an end that is induced by infinitely many singularities.

Note that all the translation surfaces in the last example have infinite area but there exist also examples for this phenomenon that have finite area: The exponential surface from Section 2.2 has one end which is induced by the unique singularity. However, a small modification by inserting a rectangle between the 0th and the -1th trapezoid yields a translation surface with two wild singularities that are inducing the same end. Furthermore, in Example 3.9 (ii) we will see a translation surface of finite area with a singular end that is not induced by any singularity.

3.2 Spaces of ends that can occur

As we have seen in Proposition 1.14, for finite translation surfaces the genus can be calculated from information on the singularities. We can ask the naive question whether the number of singularities, the number of ends, and the genus can be the only ingredients in a Gauß-Bonnet like formula for general translation surfaces. As this is not even the case for finite translation surfaces (recall that we need the multiplicities of the singularities), the answer is obviously “no”. However, it is a bit surprising how few restrictions exist for the number of singularities, the number of ends, and the genus.

Theorem 1 (No restrictions in terms of topology and number of singularities)
Let (s, g, e) be a triple of natural numbers with $s \geq 1$, $g \geq 0$, and $e \geq 1$. Then there exists a translation surface with s (nonremovable) cone angle or infinite angle singularities, genus g, and e ends.

Proof. We explicitly construct a translation surface for each triple, distinguished into several cases. The construction in the generic case consists of three steps of which the first step produces singularities, the second step produces genus and the third step produces ends. The special cases that are not covered by the generic case are treated by variations of this construction.

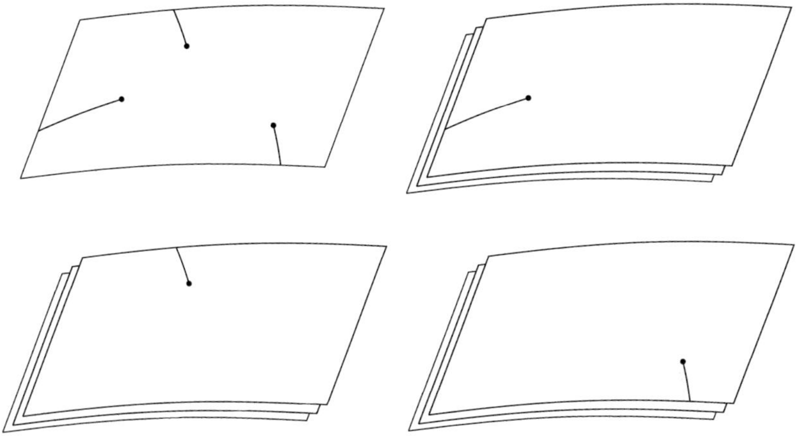

Figure 3.1: The first step of the construction produces one end and s singularities.

We start with the generic case of $s \geq 1$, $g \geq 0$, and $e \geq 3$: In the first step, consider a Euclidean plane and choose s points with norm 1. For each point consider a direction in which we can make an infinite slit so that no two slits intersect. Now for each of these points, cut open along the corresponding slit and glue in an infinite cyclic translation covering of a once-punctured Euclidean plane along a slit in the same direction (see Figure 3.1).

As a second step, consider a regular $(4g + 2)$-gon that does not intersect the slits but where one vertex is equal to one of the infinite angle singularities. Remove the interior of the $(4g + 2)$-gon. If $g = 0$, this 2-gon is simply a finite slit that extends one of the infinite slits. The edges of the $(4g + 2)$-gon come in pairs which are parallel. Glue each of these pairs except one so that two edges remain unglued (see Figure 3.2).

In the third step, cut one of the last two unglued edges into $e - 2$ segments and at each segment glue in a half-cylinder, i.e. $\mathbb{R}/k\mathbb{Z} \times [0, \infty)$ for a suitable k. Also glue in a half-cylinder at the last unglued edge (see Figure 3.3).

Then the resulting translation surface has s infinite angle singularities from the first step. Recall that one singularity originates from identifying all vertices of the regular $(4n + 2)$-gon and of the half-cylinders. It has

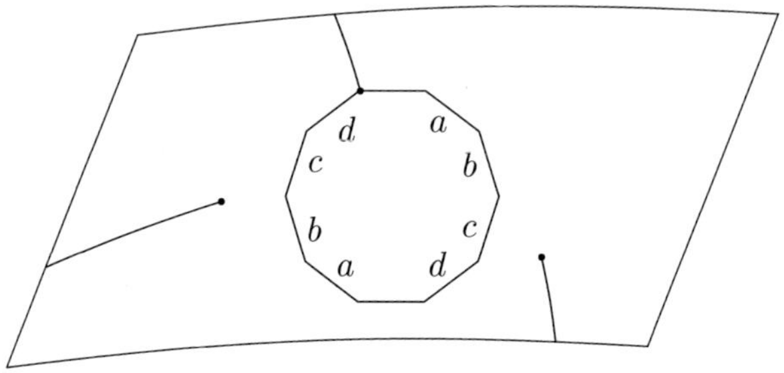

Figure 3.2: The second step of the construction produces genus g.

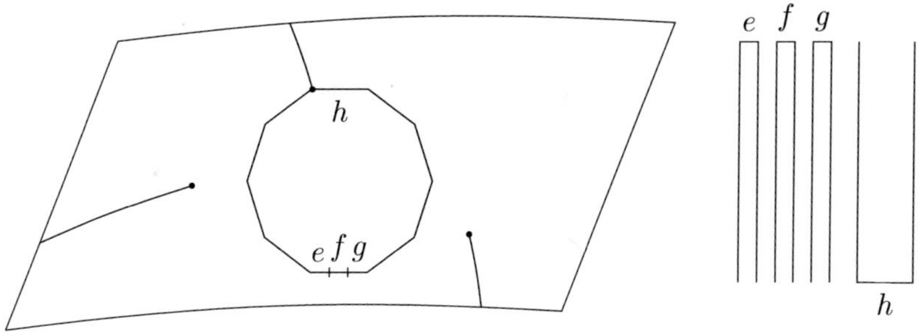

Figure 3.3: The third step of the construction produces $e-1$ ends.

one end from the first step of the construction and $e-1$ ends from the half-cylinders that are glued in. Furthermore, the genus is g because of the gluing of the $4g$ edges of the $(4g+2)$-gon.

Now we move on to the special cases, distinguished mainly by the number of ends.

If $s \geq 1$, $g \geq 1$, and $e=2$ is given then we can use the same construction as before but produce $s-1$ infinite angle singularities and one cone angle singularity in the first step. In the second step, instead of the $(4g+2)$-gon, we consider a $4g$-gon with a vertex at the cone angle singularity and with all parallel edges glued so that we can omit the third step. Then we have one end from the first step and one end induced by the puncture of the cone angle singularity.

If $s \geq 1$, $g \geq 1$, and $e=1$ is given then we can use the same construction as for $e=2$ with the $4g$-gon again but with s infinite angle singularities instead of $s-1$ infinite angle singularities and one cone angle singularity in the first step.

Now, there are still two cases missing: the case $e=2$ and $g=0$, and the case $e=1$ and $g=0$.

If $s \geq 1$, $g=0$, and $e=2$ then we produce $s-1$ infinite angle singularities and one cone angle singularity in the first step and omit the second and the third step. Hence, we have one end from the original Euclidean plane and one end induced by the cone angle singularity.

If $s \geq 1$, $g=0$, and $e=1$ then we produce s infinite angle singularities in the first step and again omit the second and the third step. □

There are two types of cases that are not covered by Theorem 1: $s=0$ and $e=0$. We know that these cases can only be realized under additional conditions.

If $s=0$ then the universal cover of the surface is the Euclidean plane. Hence, $(X, \mathcal{A})$ is either a Euclidean plane ($g=0$, $e=1$), a cylinder ($g=0$, $e=2$), or a torus ($g=1$, $e=0$).

If $e=0$ then we have a compact surface without singularities (i.e. $s=0$), so the only possibility is a torus ($g=1$).

Also if we consider infinitely many ends, there are still no restrictions in what is realizable. To show this we give the following theorem which was proven together with Camilo Ramírez Maluendas.

Theorem 2 (Existence of a translation surface with given ends) *Let E, E' be closed subsets of the Cantor set with $E' \subseteq E \neq \emptyset$. Then there exists a translation surface $(X, \mathcal{A})$ that has only cone angle singularities, i.e. the complex structure of X can be extended to $\overline{X}$, and such that* $\text{Ends}(\overline{X})$ *is homeomorphic to E and* $\text{Ends}_{nonplanar}(\overline{X})$ *is homeomorphic to E'.*

Proof. Recall from Example 1.28 that the space of ends of an infinite complete binary tree T is a Cantor set. In the same way, a closed subset $E \neq \emptyset$ of the Cantor set is homeomorphic to the space of ends of a connected subtree T_E of T. For the proof of the theorem, we will cover the tree T_E by a countable set of geodesic rays and associate to every ray a Euclidean plane. The planes are glued together dependent on the corresponding rays so that the space of ends of the resulting translation surface is homeomorphic to the space of ends of T_E.

We define the set of geodesic rays inductively: Let v_0 be the vertex in T_E which is closest to the root in T and let $\gamma_0 \colon [0, \infty) \to T_E$ be a geodesic ray starting in v_0. Now for every $n \geq 1$, let a_n be one of the edges that are closest to the root in T and whose interiors are contained in $T_E \setminus \bigcup_{i=0}^{n-1} \text{im}(\gamma_i)$ (there are at most finitely many of these edges). Let v_n be the vertex of a_n which is closer to the root in T and let $\gamma_n \colon [0, \infty) \to (T_E \setminus \bigcup_{i=0}^{n-1} \text{im}(\gamma_i)) \cup \{v_n\}$ be a geodesic ray starting in v_n (see Figure 3.4 for an example). We obtain a set $\Gamma_E = \{\gamma_n : n \in \mathbb{N}\}$ of geodesic rays in T_E so that $\cup_{\gamma \in \Gamma_E} \text{im}(\gamma) = T_E$ and so that a ray intersects the union of the previous rays exactly in its start point. Because Γ_E is ordered, it is in particular countable.

Now we construct a translation surface $(X, \mathcal{A})$ in the following way. For each geodesic ray $\gamma \colon [0, \infty) \to T_E$ in Γ_E, consider a Euclidean plane P_γ with infinitely many marked geodesic segments in horizontal direction: For each $n \in \mathbb{N}$ consider the geodesic segment s_n from $(8n, 0)$ to $(8n + 2, 0)$, the geodesic segment t_n from $(8n + 3, 0)$ to $(8n + 4, 0)$, and the geodesic segment t'_n from $(8n + 6, 0)$ to $(8n + 7, 0)$. We will consider some of the geodesic segments as slits and glue them in the following way. By the construction of Γ_E, for every ray $\gamma \in \Gamma_E \setminus \{\gamma_0\}$, there exists a $\gamma' \in \Gamma_E$ so that $\gamma(0) = \gamma'(n)$ for some $n \in \mathbb{N}$. Glue the slit s_0 in P_γ and the slit s_n in $P_{\gamma'}$. Note that there exist a $\gamma' \in \Gamma_E$ and $n \in \mathbb{N}$ with $\gamma_0(n) = \gamma'(0)$ if E has more than one element. Therefore, the union of the Euclidean planes P_γ with the described gluings is connected.

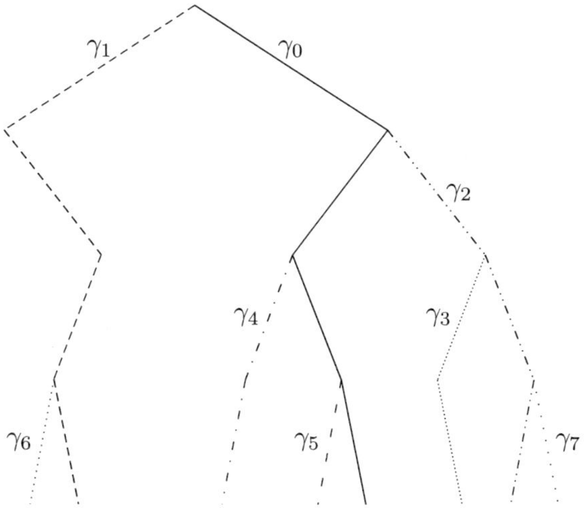

Figure 3.4: Every subtree of the infinite complete binary tree can be covered by the images of countably many rays.

Now let $e \in E$. Then there exists a unique geodesic, hence proper ray $\gamma_e : [0, \infty) \to T_E$ with $\gamma(0)$ the root of T_E and $\mathrm{end}(\gamma_e) = e$. In particular, for every $n \in \mathbb{N}$ there exists a $\gamma_{i_n} \in \Gamma_E$ and $m_n \in \mathbb{N}$ with $\gamma([n, n+1]) = \gamma_{i_n}([m_n, m_n + 1])$. Define a proper ray $r_e : [0, \infty) \to \cup_{\gamma \in \Gamma_E} P_\gamma$ in the following way: For each $n \in \mathbb{N}$, let $r_e([n, n+1])$ be a proper curve from $(8m_n + 1, 0)$ to $(8(m_n + 1) + 1, 0)$ in $P_{\gamma_{i_n}}$. If $e \in E'$ then additionally glue t_{m_n} to t'_{m_n} in $P_{\gamma_{i_n}}$ for each $n \in \mathbb{N}$.

In the constructed translation surface $(X, \mathcal{A})$, all singularities are cone angle singularities of multiplicity 2. Hence, the complex structure of X can be extended to $\overline{X}$ and $\overline{X}$ is a surface. It remains to show that $\mathrm{Ends}(\overline{X})$ is homeomorphic to E and $\mathrm{Ends}_{nonplanar}(\overline{X})$ is homeomorphic to E'.

As we have introduced before, every $e \in E$ defines a proper ray $r_e : [0, \infty) \to X$. If e' is an element of E with $e' \neq e$ then after finite time, r_e and $r_{e'}$ will never be in the same copy of the Euclidean plane again (as the rays γ_e and $\gamma_{e'}$ in T_E split up after finite time). Note that all planes P_{γ_i}, P_{γ_j} can be separated by compact sets as there is only one slit in P_{γ_i} that is glued to a slit in P_{γ_j} or to a slit in a plane which is connected to P_{γ_j} in $X \setminus P_{\gamma_i}$. Hence, $\mathrm{end}(r_e) \neq \mathrm{end}(r_{e'})$.

Now let $r\colon [0,\infty) \to X$ be a proper ray. If it is contained in one P_γ for some $\gamma \in \Gamma_E$ after finite time then $\text{end}(r) = \text{end}(r_{\text{end}(\gamma)})$ where $\text{end}(\gamma)$ is a point in E. Else, r has to leave every Euclidean plane in finite time and never come back. The only possibility to do so is to use the glued slits s_n. Each of the glued slits s_n encodes a vertex in T_E, so r uniquely defines a geodesic ray γ in T_E. Then we have $\text{end}(r) = \text{end}(r_{\text{end}(\gamma)})$.

In summary, the sets $\text{Ends}(X)$ and E are in one-to-one correspondence. Moreover, rays in T_E are close if and only if the corresponding rays in X are close: two proper rays in T_E that pass through the same edges of T_E for some time induce two proper rays in X that pass through the same copies of the Euclidean plane for some time. Therefore, $\text{Ends}(X)$ and E are homeomorphic. Furthermore, by gluing some of the slits t_n and t'_n as described, we have that $\text{Ends}_{nonplanar}(\overline{X})$ is homeomorphic to E'. □

The translation surface that arises by the construction in the proof of Theorem 2 is a branched covering of the once-punctured torus as it can be glued together from infinitely many squares. Translation surfaces of this type are called *origamis*. Finite origamis have the nice property that there exists an algorithm to determine their Veech groups (see [Sch04, Chapter 3]). For the origamis in this construction, we can check by hand that the Veech group is always

$$P = \left\{ \begin{pmatrix} 1 & t \\ 0 & s \end{pmatrix} : t \in \mathbb{R},\ s \in \mathbb{R}_+ \right\}.$$

Przytycki, Schmithüsen, and Valdez have shown that the group P is one of three possible groups that occur as an uncountable Veech group of a tame translation surface $(X, \mathcal{A})$ where $\overline{X}$ is a Loch Ness Monster (cf. [PSV11, Theorem 1.1]). The search for a translation surface with Veech group P and a given space of ends was the original starting point for the work with Camilo Ramírez Maluendas.

3.3 Relations between ends and singularities

To understand the ends of a translation surface, we are especially interested in singular ends and how they are related to the singularities of the translation surface. Nevertheless we start by proving some properties of regular ends.

Proposition 3.5 (Regular ends are planar and of infinite area) *Let $(X, \mathcal{A})$ be a translation surface. Then every regular end of $(X, \mathcal{A})$ is planar and of infinite area.*

Proof. Let e be a regular end of $(X, \mathcal{A})$ and $K \subseteq X$ a compact set so that the path-connected component C of $X \setminus K$ that defines e does not contain a singularity in its closure. This means that $\overline{C}$ is a locally flat, complete surface with boundary. If $\overline{C}$ had finite area then it would be compact which means that it cannot define an end. Hence, e has infinite area. Furthermore, C is homeomorphic to $\mathbb{R}^2$ with punctures where each puncture corresponds to a boundary component of C or to an end which is not a point in the metric completion. This means that C has genus 0 and hence, e is a planar end. □

For singular ends, there are much more possibilities. First of all, some singular ends are induced by a singularity, some are not (see Example 3.4). Furthermore, a singular end can be planar or nonplanar and can be of finite area or of infinite area. Even more, neither of these four properties implies one of the others in this case. We give examples for all four combinations by the next proposition and the subsequent example.

Proposition 3.6 (Ends induced by cone angle singularities) *Let $(X, \mathcal{A})$ be a translation surface and σ a cone angle singularity. Then the induced end* $\mathrm{end}(\sigma)$ *can not be induced by another singularity. Moreover,* $\mathrm{end}(\sigma)$ *is planar and of finite area.*

Proof. Let $\epsilon > 0$ small enough so that there exists a finite cyclic translation covering $B(\sigma, 3\epsilon) \setminus \{\sigma\} \to B(0, 3\epsilon) \setminus \{0\} \subseteq \mathbb{R}^2$. This shows that $K := \overline{B(\sigma, 2\epsilon)} \setminus B(\sigma, \epsilon)$ is compact as a closed annulus in $\mathbb{R}^2$ is compact.

Suppose there exists a singularity σ' with $\mathrm{end}(\sigma') = \mathrm{end}(\sigma)$. Then there exist proper rays $r, r' \colon [0, \infty) \to X$ with $\mathrm{end}(r) = \mathrm{end}(\sigma)$ and $\mathrm{end}(r') = \mathrm{end}(\sigma')$. Furthermore, there exists an $n \in \mathbb{N}$ so that $r([n, \infty))$ and $r'([n, \infty))$ are contained in the same path-connected component of $X \setminus K$. As $r([n, \infty))$ is contained in $B(\sigma, \epsilon) \setminus \{\sigma\}$, the same is true for $r'([n, \infty))$. Hence, the distance between σ and σ' is less than ϵ and by the choice of ϵ we have $\sigma' = \sigma$.

As σ is a cone angle singularity, there exists a finite translation covering $B(\sigma, \epsilon) \setminus \{\sigma\} \to B(0, \epsilon) \setminus \{0\}$. Therefore, $B(\sigma, \epsilon) \setminus \{\sigma\}$ is planar and of finite area and we can deduce that $\mathrm{end}(\sigma)$ is planar and of finite area. □

The argument in the proof of Proposition 3.6 does not work for a wild singularity σ as it is not true in general that $\overline{B(\sigma, 2\epsilon)} \setminus B(\sigma, \epsilon)$ is compact. For instance, we will see in Section 4.1 that for the singularity of the exponential surface this set decomposes into infinitely many connected components, hence it can not be compact.

If σ is an infinite angle singularity of a translation surface $(X, \mathcal{A})$ then $\text{end}(\sigma)$ can also be induced by other singularities as we have seen in Example 3.4 (ii). The set $\overline{B(\sigma, 2\epsilon)} \setminus B(\sigma, \epsilon)$ in the proof of Proposition 3.6 can not be compact for an infinite angle singularity. In fact, $\text{end}(\sigma)$ is of infinite area because for every compact set $K \subseteq X$, there exists an ϵ-neighborhood of σ which is disjoint to K and has infinite area. This gives the first of the following examples.

Example 3.7 (Singular ends can be of different types).

(i) Let $(X, \mathcal{A})$ be an infinite cyclic translation covering of a once-punctured Euclidean plane. Then $(X, \mathcal{A})$ has one infinite angle singularity σ and one end which is induced by σ. We have that $\text{end}(\sigma)$ is planar and of infinite area.

(ii) For the exponential surface in Section 2.2, the unique end is induced by the unique singularity. Furthermore, the end is nonplanar and of finite area as every unbounded path-connected component of the complement of a compact set has infinite genus and finite area.

(iii) We have seen in Example 3.4 (iii) for the case $n = 1$ that for the harmonic series decoration on a Euclidean plane as in Section 2.4, again the unique end is induced by the unique singularity. Furthermore, the end is nonplanar and of infinite area as every unbounded path-connected component of the complement of a compact set has infinite genus and infinite area.

(iv) We can modify the previous example by choosing the lengths of the segments not to be shrinking but by choosing all segments to be of the same length. Then the unique singularity is an infinite angle singularity but it is still inducing the unique end and the end is nonplanar and of infinite area.

In view of Proposition 3.6, we can ask whether a singular end which is planar and of finite area has to be induced by a cone angle singularity.

The following proposition and the subsequent example show that a weaker statement is true and that both conditions on the end, planar and of finite area, are needed for the conclusion.

Proposition 3.8 (Weakened version of converse statement of Proposition 3.6)
Let $(X, \mathcal{A})$ be a translation surface with finitely many singularities and e a singular, planar end of $(X, \mathcal{A})$ of finite area. Then there exists a singularity σ such that $\text{end}(\sigma) = e$.

Proof. Let $r\colon [0, \infty) \to X$ be a proper ray so that $\text{end}(r) = e$. Furthermore, let $K \subseteq X$ be a compact set so that the path-connected component C of $X \setminus K$ which contains $r([n, \infty))$ for an $n \in \mathbb{N}$ is planar and has finite area. Let $\overline{C}$ be the closure of C in $\overline{X}$ and Σ_C the set of singularities that are contained in $\overline{C} \setminus K$. As e is singular, this set is not empty.

For every $\sigma \in \Sigma_C$ choose $\epsilon_\sigma > 0$ so that $\overline{B(\sigma, 3\epsilon_\sigma)} \setminus \{\sigma\} \subseteq C$. Because C is planar and has finite area, $\overline{B(\sigma, 2\epsilon_\sigma)} \setminus B(\sigma, \epsilon_\sigma)$ is planar and has finite area for every $\sigma \in \Sigma_C$. As this set is also complete, it is compact. Let

$$K' := K \cup \bigcup_{\sigma \in \Sigma_C} \left(\overline{B(\sigma, 2\epsilon_\sigma)} \setminus B(\sigma, \epsilon_\sigma)\right)$$

and C' the path-connected component of $X \setminus K'$ with $r([n', \infty)) \subseteq C'$ for an $n' \in \mathbb{N}$.

As r defines a singular end, C' is equal to $B(\sigma, \epsilon_\sigma) \setminus \{\sigma\}$ for a $\sigma \in \Sigma_C$. Hence, $e = \text{end}(r) = \text{end}(\sigma)$. ☐

We will now show in an example that both conditions on the end, planar and of finite area, are needed for the conclusion in Proposition 3.8. The example also indicates that the ends of singularities can converge to another end. The limit is a singular end but it does not have to be induced by a singularity.

Example 3.9 (Singular ends that are not induced by a singularity).

(i) For every pair $(n, m) \in \mathbb{Z} \times \mathbb{Z} \setminus \{(n, 0) : n \neq 0\}$, consider a copy $P_{n,m}$ of a Euclidean plane. On the Euclidean plane $P_{0,0}$ consider slits $s_{n,0}$ from $(0, n)$ to $(1, n)$ for each $n \in \mathbb{Z}$. For each $P_{n,m}$ with $(n, m) \neq (0, 0)$ consider a slit $s_{n,m}$ from $(0, 0)$ to $(1, 0)$ (see Figure 3.5).

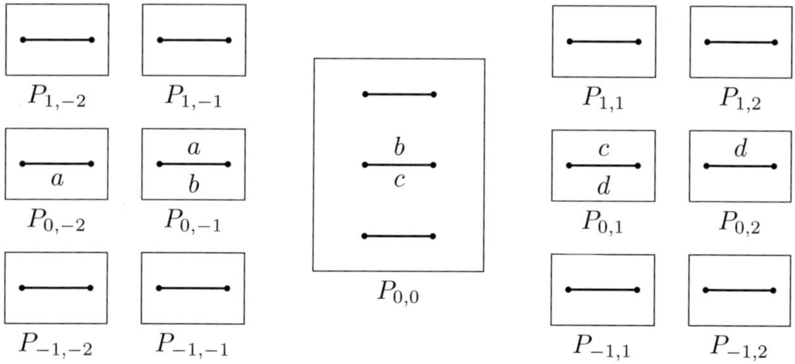

Figure 3.5: When gluing all slits in the same way as indicated for one family, we obtain a translation surface which has a planar, singular end that is not induced by a singularity.

By gluing the lower part of $s_{n,m}$ to the upper part of $s_{n,m+1}$, we obtain a translation surface $(X, \mathcal{A})$ with genus 0. This implies that every end of $(X, \mathcal{A})$ is a planar end. The translation surface has two infinite angle singularities $\sigma_{n,left}$ and $\sigma_{n,right}$ for each family of slits $\{s_{n,m} : m \in \mathbb{Z}\}$ with $n \in \mathbb{Z}$. These two singularities $\sigma_{n,left}$ and $\sigma_{n,right}$ induce the same end for every $n \in \mathbb{Z}$.

Moreover, there exists a regular end for each $P_{n,m}$ except for $P_{0,0}$: consider for example the path-connected components of the complement of the compact set $\overline{B(0,3)} \setminus B(0,2)$ in $P_{n,m}$.

For $P_{0,0}$ the proper ray $r\colon [0, \infty) \to P_{0,0}$, $x \mapsto (-1, x)$ defines a singular end. However, for every $N \geq 1$, we have the compact set $\{(x, y) \in P_{0,0} : \max\{x, y\} = N + \frac{1}{2}\}$ which separates end(r) from the ends of $\sigma_{n,left}$ and $\sigma_{n,right}$ for $n \in [-N, N]$. This means that end(r) is not induced by a singularity.

(ii) Consider the union of the rectangles R_n from $(n, 0)$ to $(n+1, 2^{-|n|})$ for every $n \in \mathbb{Z}$. By gluing opposite edges of this figure we obtain a translation surface $(X, \mathcal{A})$ (see Figure 3.6).

The translation surface $(X, \mathcal{A})$ has infinitely many cone angle singularities of multiplicity 3 and no other singularities. As the area of

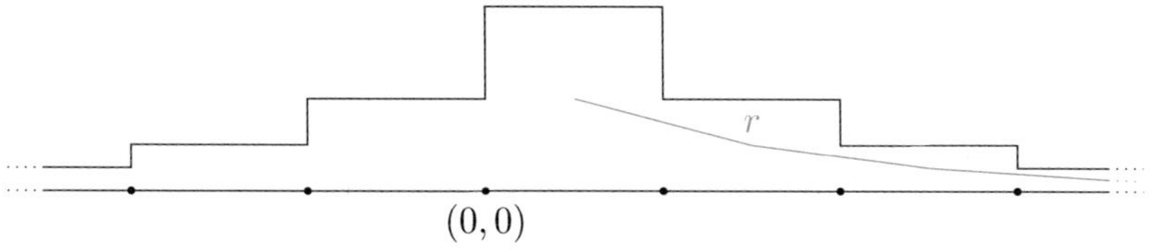

Figure 3.6: When gluing opposite horizontal and vertical segments, we obtain a translation surface so that end(r) is a singular end of finite area which is not induced by a singularity.

the nth rectangle is $2^{-|n|}$, the area of X is finite and hence every end of $(X, \mathcal{A})$ is of finite area.

For each cone angle singularity there exists an induced end. Now let $r\colon [0, \infty) \to X$ be the proper ray which goes to the right in a way so that $r([n, n+1])$ is the geodesic segment from $(n + \frac{1}{2}, 2^{-n-1})$ to $(n + \frac{3}{2}, 2^{-n-2})$. Then r defines an end which is singular but is not induced by a singularity.

The last example also shows that the space of singularities and the space of ends do not have to be in one-to-one correspondence, even if the translation surface has finite area. To complete this section, we exhibit that there exists a one-to-one correspondence between the space of singularities and the space of ends if the translation surface does not only have finite area but the metric completion is compact.

Proposition 3.10 (Correspondence between ends and singularities for compact metric completion)
Let $(X, \mathcal{A})$ be a translation surface so that the metric completion $\overline{X}$ is compact. Then the space of ends is finite and in one-to-one-correspondence to the set of singularities.

Proof. The number of singularities of $(X, \mathcal{A})$ is finite as it is a discrete closed subset of the compact space $\overline{X}$. We show that no two singularities induce the same end and that every end is induced by a singularity.

Let σ be a singularity and $\epsilon_\sigma > 0$ so that the distance between σ and every other singularity is greater than $3\epsilon_\sigma$. As $K := \overline{B(\sigma, 2\epsilon_\sigma)} \setminus B(\sigma, \epsilon_\sigma)$ is closed and $\overline{X}$ is compact, K is also compact. Moreover, K is contained in X and $B(\sigma, \epsilon_\sigma)$ does not contain any other singularity than σ. As in the proof of Proposition 3.6, this shows that no other singularity can induce $\text{end}(\sigma)$.

As $\overline{X}$ has finite area, every end has finite area so it is singular. We can show in the same way as in the proof of Proposition 3.8 that every end is induced by a singularity: The condition in Proposition 3.8 that the end is planar is replaced by the condition that $\overline{X}$ is compact, so $\overline{B(\sigma, 2\epsilon_\sigma)} \setminus B(\sigma, \epsilon_\sigma)$ is again compact for every singularity σ and ϵ_σ as before.

Altogether we have a map $\overline{X} \setminus X \to \text{Ends}(X)$, $\sigma \mapsto \text{end}(\sigma)$ which is bijective. As both spaces carry the discrete topology, the map is actually a homeomorphism. □

Note that it is crucial for the proof of the previous proposition that we concluded the convention at the beginning of Section 1.2 that the set of singularities is discrete. When we consider the excluded translation surface that is defined by an open disk, we see that the combination of one end and infinitely many singularities is possible, even if the metric completion is compact.

We summarize the results of this section in the following theorem.

Theorem 3 (Relations between ends and singularities)
Let $(X, \mathcal{A})$ be a translation surface, e an end of $(X, \mathcal{A})$ and σ a singularity of $(X, \mathcal{A})$.

(i) If e is a regular end then it is planar and of finite area.

(ii) If e is a singular end then it can be induced by every number of singularities, including zero.

(iii) If σ is a cone angle singularity then $\text{end}(\sigma)$ *is planar and of finite area.*

(iv) If σ is an infinite angle singularity then $\text{end}(\sigma)$ *is of infinite area and can be planar or nonplanar.*

(v) If $(X, \mathcal{A})$ has finitely many singularities and e is a singular, planar end of finite area then e is induced by a singularity.

(vi) If $\overline{X}$ is compact then the ends and the singularities are in one-to-one-correspondence.

4 Around wild singularities

Our aim is to understand the topology and geometry of a translation surface. Locally, this is only interesting for singularities as every regular point has a neighborhood which is flat. For a cone angle or infinite angle singularity and for an $\epsilon > 0$ that is small enough, we know exactly what the ϵ-neighborhood of the singularity looks like by Definition 1.12.

For a wild singularity, everything seems possible for an ϵ-neighborhood at the first glance. In the beginning of this chapter, we verify for some basic properties whether they can or cannot occur for ϵ-neighborhoods of wild singularities. Then we move on to saddle connections, in particular to saddle connections that are short enough so that their images are contained in an ϵ-neighborhood of the singularity. In Chapter 5 we will use the results from this chapter to show that the existence of a wild singularity even encodes information on the genus of the translation surface.

4.1 Topological and geometric properties of a neighborhood of a singularity

We start by checking whether the idea is correct that an ϵ-neighborhood of a singularity is a somehow deformed once-punctured disk. The discussion consists of a positive result and several examples of what can go wrong for wild singularities. The positive statement is that every ϵ-neighborhood of a wild singularity is path-connected (previously published as [Ran14, Proposition 12]).

Proposition 4.1 (Balls around singularities are path-connected) *Let $(X, \mathcal{A})$ be a translation surface, σ a singularity, and $\epsilon > 0$. Then $B(\sigma, \epsilon) \cap X$ is path-connected.*

Proof. For a fixed $\epsilon > 0$, let $x, y \in B(\sigma, \epsilon) \cap X$ and $\delta > 0$ small enough so that $d(x, \sigma) < \epsilon - 3\delta$ and $d(y, \sigma) < \epsilon - 3\delta$. Furthermore, let $(z_n)_{n\in\mathbb{N}} \subseteq X$

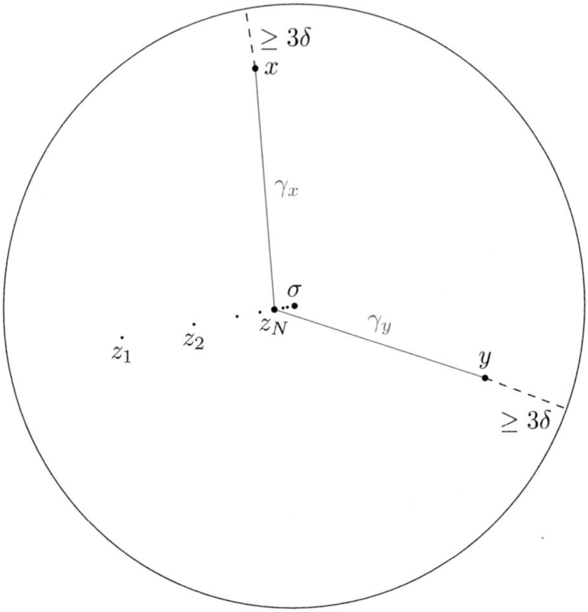

Figure 4.1: The ϵ-neighborhood of a singularity σ and a Cauchy sequence (z_n) converging to σ.

be a Cauchy sequence converging to σ and $N \in \mathbb{N}$ with $d(\sigma, z_N) < \delta$ (see Figure 4.1). Then we have

$$d(x, z_N) \leq d(x, \sigma) + d(\sigma, z_N) < \epsilon - 2\delta$$

and $d(y, z_N) < \epsilon - 2\delta$.

As the metric of X is the path length metric, there exists a curve γ_x in X which connects x to z_N and has at most length $\epsilon - \delta$. This means that for every point in the image of γ_x the distance to σ is at most $d(\sigma, z_N) + (\epsilon - \delta) < \epsilon$. Hence, γ_x is a curve in $B(\sigma, \epsilon) \cap X$.

In the same way we can define a curve γ_y in $B(\sigma, \epsilon) \cap X$ from y to z_N and by concatenation of γ_x with the reversed curve of γ_y we obtain a curve from x to y in $B(\sigma, \epsilon) \cap X$. This means that $B(\sigma, \epsilon) \cap X$ is path-connected. $\square$

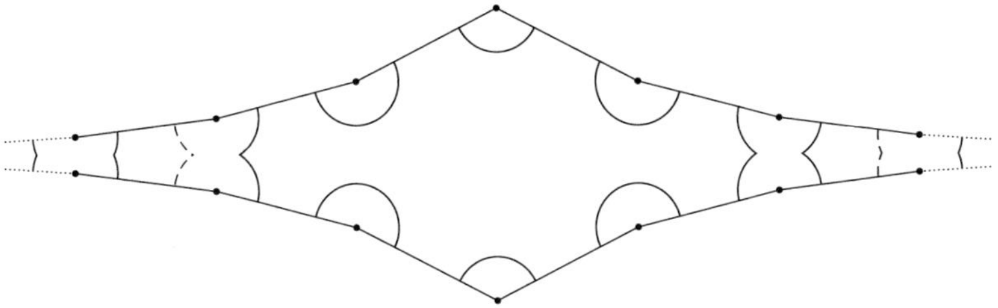

Figure 4.2: The boundary of the $\frac{1}{3}$-neighborhood for the wild singularity of the exponential surface: the dashed curve is one boundary component of the neighborhood.

The path-connectedness is also fulfilled for $B(\sigma,\epsilon) \subseteq \overline{X}$. The proof is literally the same as the metric of $\overline{X}$ is also the path length metric.

So, a crucial property of a disk is fulfilled. However, it can happen for a wild singularity σ that the boundary of $B(\sigma,\epsilon)$ consists of more than one component. Consider for example the exponential surface in Section 2.2 and an $\epsilon \in [2^{-n}, 2^{-n+1})$ for an $n \geq 2$. Then for every trapezoid (except for the first $n+1$ trapezoids in both directions) the bases of the trapezoid are contained in the ϵ-neighborhood of σ. However, no trapezoid is completely contained in $B(\sigma,\epsilon)$, so there are two pieces of boundary in each trapezoid (see Figure 4.2). These pieces come in pairs so that they form closed curves. Hence, we have infinitely many boundary components of the ϵ-neighborhood of σ for every $\epsilon \in (0, \frac{1}{2})$.

This example also shows that the number of boundary components of an ϵ-neighborhood is not directly related to the number of rotational components: for the exponential surface, we have infinitely many boundary components of an ϵ-neighborhood and two rotational components. The other way around, for the Chamanara surface, we have one boundary component of the ϵ-neighborhood for each $\epsilon \in (0, \frac{1}{2})$ and infinitely many rotational components.

Moreover, we observe that $B(\sigma, 2\epsilon) \setminus \overline{B(\sigma,\epsilon)} \subseteq X$ need not be an annulus (see Figure 4.3). For the exponential surface and an $\epsilon \in (0, \frac{1}{2})$, the set is not even connected, in fact it decomposes into infinitely many connected components. And as there exists an infinite subset of the connected components so that the distance of each two connected components is at

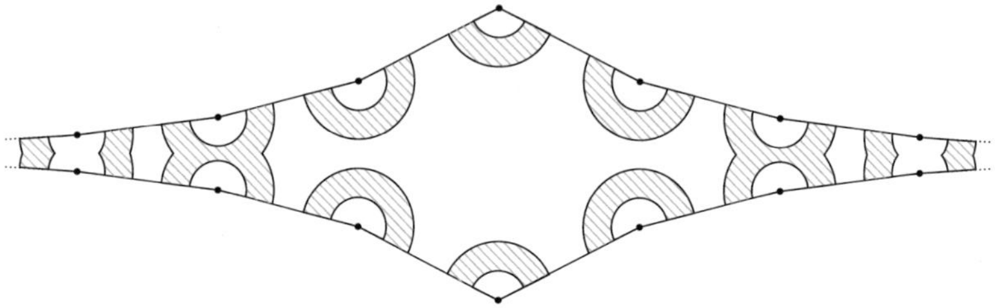

Figure 4.3: The set $\overline{B(\sigma, \frac{2}{5})} \setminus B(\sigma, \frac{1}{5})$ for the wild singularity of the exponential surface.

least 1, the closed set $\overline{B(\sigma, 2\epsilon)} \setminus B(\sigma, \epsilon)$ has finite area but is not compact. As was indicated in the remark following Proposition 3.6, this means that the set $\overline{B(\sigma, 2\epsilon)} \setminus B(\sigma, \epsilon)$ can not be used to distinguish end(σ) from other ends of the translation surface.

Let us now turn to the area of a neighborhood. For a cone angle singularity σ of multiplicity k, the ϵ-neighborhood $B(\sigma, \epsilon)$ has area $k \cdot \epsilon^2 \cdot \pi$. The same calculation also shows that the area of an ϵ-neighborhood of an infinite angle singularity is infinite. For a wild singularity, the area of an ϵ-neighborhood can be finite or infinite. Every translation surface with finite area and a wild singularity can serve as an example for a neighborhood that has finite area, for instance the Chamanara surface and the exponential surface. For a Euclidean plane with a harmonic series decoration as in Section 2.4, the ϵ-neighborhood of the wild singularity σ has infinite area as the infinite strip

$$\left(\sum_{k=1}^{\lceil \frac{1}{\epsilon} \rceil} \frac{1}{k}, \infty\right) \times \left[-\frac{\epsilon}{2}, \frac{\epsilon}{2}\right]$$

is contained in $B(\sigma, \epsilon)$.

For a cone angle or infinite angle singularity σ and an $\epsilon > 0$ small enough, the ϵ-neighborhood $B(\sigma, \epsilon)$ of σ in $\overline{X}$ is simply connected. This statement is not true for a wild singularity σ as we can see in Figure 4.2.

In fact, we will show in Proposition 4.4 that for every wild singularity σ and every $\epsilon > 0$ there exists a saddle connection from σ to σ whose image is contained in $B(\sigma, \epsilon)$. A saddle connection can never be null-homotopic as it would bound a locally flat disk and contracting the saddle connection would yield a homotopic curve which has smaller length. The existence of this homotopic curve is impossible as a saddle connection is a geodesic and a geodesic is minimal in its homotopy class. Hence, the existence of a saddle connection from σ to σ whose image is contained in $B(\sigma, \epsilon)$ implies that $B(\sigma, \epsilon)$ cannot be simply connected.

4.2 Saddle connections in a dense set of directions

As we have indicated in the last section, saddle connections contain information on the topology of the neighborhood of a singularity. Therefore, we continue our study of ϵ-neighborhoods of singularities by investigating saddle connections.

It is proven by Masur in [Mas86, Theorem 2] that for finite translation surfaces the directions of saddle connections are dense in S^1. We show that a similar result is true for translation surfaces with finite area. For this, we construe saddle connections as linear approaches in the following way. Recall that for a translation surface $(X, \mathcal{A})$, every saddle connection $\gamma\colon [0, l] \to \overline{X}$ is geodesic and its image has an orientation by definition. So the curve $\gamma_{|(0,l)}$ and its reversed curve define two linear approaches, one to the singularity $\gamma(0)$ and one to the singularity $\gamma(l)$. We call the first linear approach $[\gamma_+]$ and the second one $[\gamma_-]$.

We use this perception of saddle connections to formulate our statement on directions of saddle connections as a corollary of the following proposition.

Proposition 4.2 (Saddle connections are dense in $\mathcal{L}(\sigma)$)
Let $(X, \mathcal{A})$ be a translation surface of finite area and σ a singularity. Furthermore, let $[\gamma] \in \mathcal{L}(\sigma)$ be a linear approach which is contained in the interior of the rotational component $\overline{[\gamma]}$. Then every neighborhood of $[\gamma]$ in $\mathcal{L}(\sigma)$ contains a linear approach defined by a saddle connection. In particular, the linear approach can be chosen to be contained in $\overline{[\gamma]}$.

Proof. Let γ be a representative of $[\gamma]$, $\theta \in S^1 = \mathbb{R}/2\pi\mathbb{Z}$ the direction of $[\gamma]$, and $((-y, y), \epsilon, i_\epsilon)$ an angular sector with $f_{((-y,y),\epsilon,i_\epsilon)}(0) = [\gamma]$. Furthermore, let $t > 0$ and $r > 0$ be small enough so that $B(\gamma(t), r)$ is

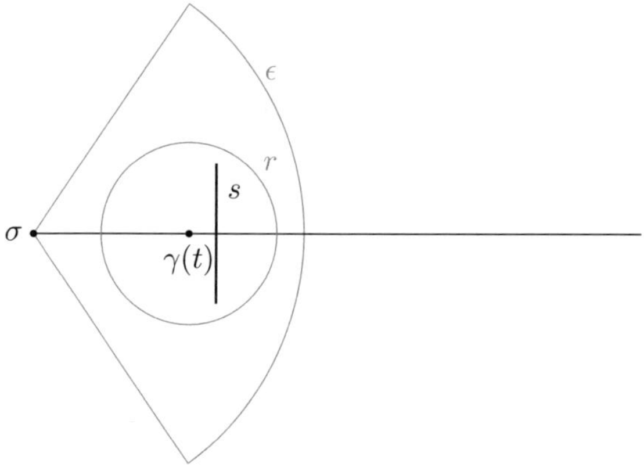

Figure 4.4: Choice of a segment s in $B(\gamma(t), r)$ which is perpendicular and symmetric to $\mathrm{im}(\gamma)$.

contained in $\mathrm{im}(i_\epsilon) \subseteq X$. We show that there exists a linear approach in $B(\gamma, t, r)$ defined by a saddle connection.

Choose a geodesic segment s in $B(\gamma(t), r)$ which is perpendicular and symmetric to $\mathrm{im}(\gamma)$ (see Figure 4.4). Consider the images of s under the geodesic flow F_θ for all times until it hits a singularity for some point in s. This has to happen after finite time because X has finite area: If the images of s under F_θ would not hit a singularity, it would be defined for all time and then by Poincaré recurrence (see Proposition 1.41) it follows that σ is hit by the flowed s.

The closure of the union of $\mathrm{im}(i_\epsilon)$ and the rectangle swept out by the geodesic segment s under the geodesic flow F_θ is a star domain with respect to the base point of the angular sector. Therefore, there exists a geodesic segment from σ to the hit singularity which is completely contained in the union of $\mathrm{im}(i_\epsilon)$ and the rectangle swept out by s (see Figure 4.5). This geodesic segment is a saddle connection and the linear approach which belongs to the saddle connection is contained in $B(\gamma, t, r)$. As the linear approach is also contained in the same rotational component as $[\gamma]$, we found a saddle connection as desired. □

We can now show that for every given direction there exists a saddle connection whose direction is close to the given one.

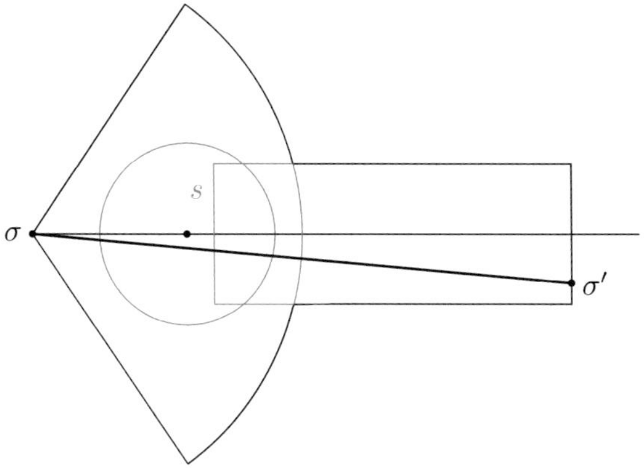

Figure 4.5: The geodesic from σ to σ' is a saddle connection for which one of the corresponding linear approaches is contained in $B(\gamma, t, r)$.

Corollary 4.3 (Set of directions of saddle connection is dense in S^1). Let $(X, \mathcal{A})$ be a translation surface with finite area and at least one singularity σ. Then the set of directions of saddle connections is dense in $S^1 = \mathbb{R}/2\pi\mathbb{Z}$.

Proof. The singularity σ has a rotational component c of length at least π. To see this, choose $\epsilon > 0$ so that $B(\sigma, 2\epsilon) \cap (\overline{X} \setminus X) = \{\sigma\}$. For a regular point $x \in B(\sigma, \epsilon)$, we have $d := d(x, \overline{X} \setminus X) = d(x, \sigma) > 0$ and that $B(x, d) \subseteq X$ is a locally flat disk with $\sigma \in \overline{B(x, d)} \subseteq \overline{X}$. Then the geodesic segment $\gamma\colon (0, d) \to B(x, d)$ which connects σ to x defines a linear approach $[\gamma] \in \mathcal{L}(\sigma)$ that is contained in a rotational component c of length at least π.

Let $\theta \in S^1$, $[\gamma]$ a linear approach contained in c with direction θ or $-\theta$, and γ a representative of $[\gamma]$. Then the direction of the saddle connection that we obtain from the proof of Proposition 4.2 differs at most $\arcsin(\frac{r}{t})$ from θ with $t > 0$, $r > 0$ as in the proof of Proposition 4.2. By making r smaller, we can lessen the difference and hence obtain a saddle connection with direction as close to θ as desired.

Note that this saddle connection also works for the opposite direction as we can consider the reversed curve of the saddle connection. $\square$

4.3 Short saddle connections and xossiness

For tame translation surfaces with finitely many singularities, the lengths of the saddle connections are bounded from below. This is because for every singularity σ there exists an $\epsilon_\sigma > 0$ so that $B(\sigma, \epsilon_\sigma) \setminus \{\sigma\}$ is locally flat. Therefore the lengths of the saddle connections are bounded from below by $\min\{\epsilon_\sigma : \sigma \text{ singularity}, B(\sigma, \epsilon_\sigma) \setminus \{\sigma\} \text{ locally flat}\}$.

For wild singularities, this argument does not work. In fact, for a wild singularity, the opposite is true.

Proposition 4.4 (Existence of short saddle connections) *Let $(X, \mathcal{A})$ be a translation surface and σ a wild singularity. Then for every $\epsilon > 0$, there exists a saddle connection connecting σ to itself of length less than ϵ.*

Proof. As the singularities are discrete by the convention in Section 1.2, there exists an $\epsilon' > 0$ so that σ is the only singularity in $B(\sigma, \epsilon') \subseteq \overline{X}$. We distinguish between the following two cases.

Case 1: There exists a rotational component which is bounded in at least one direction. This means, σ itself is an obstacle to extending the rotational component in that direction. Hence, there exist curves from σ to σ shorter than any given length, for instance shorter than ϵ. In $\overline{X}$, there exists a geodesic which is homotopic to such a curve (see [BH99, Part II, Theorem 4.13]). This geodesic in $\overline{X}$ is a chain of saddle connections connecting σ to itself and each of the saddle connections has length less than ϵ.

Case 2: Every rotational component is unbounded. Assume there exists a minimal length $\epsilon < \epsilon'$ of saddle connections from σ to σ. Then for every linear approach $[\gamma]$ to σ, there exists a representative $\gamma \in \mathcal{L}^\epsilon(\sigma)$ as otherwise it would belong to a saddle connection of length less than ϵ. Therefore, there exists a cyclic translation covering from $B(\sigma, \epsilon) \setminus \{\sigma\}$ to the once-punctured disk $B(0, \epsilon) \setminus \{0\} \subseteq \mathbb{R}^2$. This means that σ is not a wild singularity. □

Even if there exist arbitrarily short saddle connections, a closed geodesic in X can not be intersected by arbitrarily short saddle connections as we will see in the definition below. We can give a lower bound on the length of intersecting saddle connections by the *immersion radius* of a closed geodesic (previously defined in [Ran14, Definition 17]).

The immersion radius of a point or of a curve is defined similar to the well-known *injectivity radius*. In contrast to the injectivity radius, we allow that the image of the disk that we map into X overlaps itself.

Definition 4.5 (Immersion radius)
Let $(X, \mathcal{A})$ be a translation surface.

(i) For a regular point $x \in X$, we define the *immersion radius* $\mathrm{ir}(x)$ by

$$\mathrm{ir}(x) := d(x, \overline{X} \setminus X) \in (0, \infty].$$

This is well-defined as $\overline{X}$ is a metric space and the set of singularities $\overline{X} \setminus X$ is discrete and hence closed in $\overline{X}$. Note that the open $\mathrm{ir}(x)$-neighborhood of x does not contain a singularity but its closure does (if $(X, \mathcal{A})$ has at least one singularity).

(ii) For a curve $\gamma\colon [0, l] \to X$, we define the *immersion radius* $\mathrm{ir}(\gamma)$ by

$$\mathrm{ir}(\gamma) := \inf\{\mathrm{ir}(\gamma(t)) : t \in [0, l]\}.$$

As the image of γ is compact, we can cover it by finitely many disks $B(\gamma(t_i), \mathrm{ir}(\gamma(t_i)))$ around points $\gamma(t_i)$ on the curve. Then the set $\partial\big(\bigcup B(\gamma(t_i), \mathrm{ir}(\gamma(t_i)))\big)$ is compact and does not intersect $\mathrm{im}(\gamma)$, thus its distance to $\mathrm{im}(\gamma)$ is positive. As this distance is a lower bound for the immersion radius of the curve, this means that $\mathrm{ir}(\gamma) > 0$ still holds and again the open $\mathrm{ir}(\gamma)$-neighborhood of $\mathrm{im}(\gamma)$ is locally flat. In particular, every saddle connection that intersects γ has at least length $2\,\mathrm{ir}(\gamma) > \mathrm{ir}(\gamma)$.

(iii) A saddle connection can be seen either as a curve in $\overline{X}$ or as a geodesic defined on an open interval. Because of that we have to slightly modify the notion of immersion radius, compared to curves in X as before. As this modification only works for saddle connections γ for which both corresponding linear approaches $[\gamma_+]$ and $[\gamma_-]$ belong to a rotational component that is large enough, we restrict on the following type of saddle connections:

Let σ_+, σ_- be two (not necessarily different) singularities of $(X, \mathcal{A})$ and $\gamma\colon [0, l] \to \overline{X}$ a saddle connection from σ_+ to σ_-. Let $[\gamma_+]$ and $[\gamma_-]$ be linear approaches as in the remark preceding Proposition 4.2

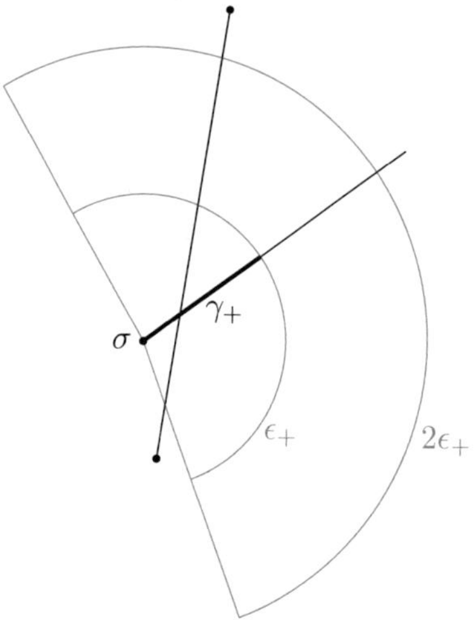

Figure 4.6: Any saddle connection that intersects the well-immersed saddle connection γ close to its start point has at least length ϵ_+.

on page 91. We say that γ is a *well-immersed* saddle connection if there exists an angular sector $((0,\pi), 2\epsilon_+, i_{2\epsilon_+})$ with base point σ_+ so that $[\gamma_+] \in \operatorname{im}(f_{((0,\pi),2\epsilon_+,i_{2\epsilon_+})})$ and the same is true for $[\gamma_-]$ and some angular sector $((0,\pi), 2\epsilon_-, i_{2\epsilon_-})$.

Now let $\gamma\colon [0,l] \to \overline{X}$ be a well-immersed saddle connection from σ_+ to σ_- and let γ_+ be the representative of $[\gamma_+]$ in $\mathcal{L}^{\epsilon_+}(\sigma_+)$. Because the length of $[\gamma_+]$ is at least π, the image of γ_+ is contained in a locally flat, open half-disk (see Figure 4.6). Suppose there exists a saddle connection that intersects γ_+. Then it has to start outside of the image of $i_{2\epsilon_+}$ and go through the half-annulus-like set

$$\left(i_{2\epsilon_+} \circ f\right)\left(\{z \in \mathbb{C} : \log \epsilon_+ \leq \operatorname{Re}(z) < \log 2\epsilon_+, \operatorname{Im}(z) \in (0,\pi)\}\right)$$

to intersect the image of γ_+. This means that the intersecting saddle connection has at least length ϵ_+.

A similar statement holds true for γ_- and ϵ_-. Then $\gamma_c := \gamma_{|[\epsilon_+, l-\epsilon_-]}$ is a curve as in (ii) with a compact image and a well-defined immersion radius $\epsilon_c := \operatorname{ir}(\gamma_c) > 0$.

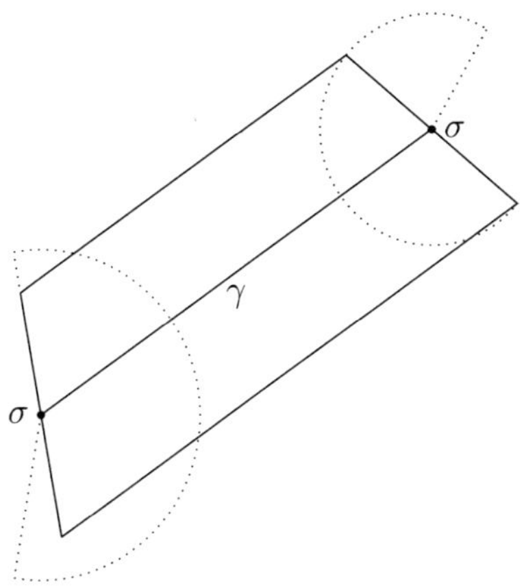

Figure 4.7: An example of a trapezoidal neighborhood of a saddle connection γ.

In particular, $\min\{\epsilon_+, \epsilon_c, \epsilon_-\}$ is a lower bound for the length of saddle connections that intersect γ. We define the *(generalized) immersion radius* $\mathrm{ir}(\gamma)$ by

$$\mathrm{ir}(\gamma) := \sup\{\min\{\epsilon_+, \epsilon_c, \epsilon_-\} : \epsilon_+, \epsilon_- \text{ small enough so that angular sectors as described above exist}\}.$$

In the case of a regular point $x \in X$, the term "immersion radius" is reasonable as an open disk of radius $\mathrm{ir}(x)$ can be immersed and the image under the immersion is a locally flat neighborhood of the point x. Similarly, for a closed geodesic γ, an open cylinder of height $2 \cdot \mathrm{ir}(\gamma)$ and circumference the length of γ can be immersed and the image is a locally flat tubular neighborhood of the curve γ.

In the case of well-immersed saddle connections, we can immerse an open trapezoid of height $2 \cdot \mathrm{ir}(\gamma)$ so that the median has the same length as the saddle connection (see Figure 4.7). The image forms a neighborhood of the interior of the saddle connection and around the singularities we have the images of angular sectors of length π.

Note that for a rotational component of length strictly greater than π, there always exists an angular sector $((0, \pi), \epsilon, i_\epsilon)$. This is true because the length of the longest representative of a linear approach varies lower semi-continuously in a rotational component and hence has a minimum on a compact set of linear approaches (see [BV13, Corollary 2.7 and the subsequent remark]). Hence, if for a saddle connection γ the linear approaches $[\gamma_+]$ and $[\gamma_-]$ are contained in rotational components of length strictly greater than π then γ is well-immersed.

Lemma 4.6 (Immersion radius is Lipschitz continuous). For a translation surface $(X, \mathcal{A})$, the map $\operatorname{ir}\colon X \to (0, \infty]$, $x \mapsto \operatorname{ir}(x)$ is Lipschitz continuous with constant 1. In particular, the map is continuous.

Proof. For all $x_1, x_2 \in X$ and $\sigma \in \overline{X} \setminus X$, we have

$$\operatorname{ir}(x_1) = d(x_1, \overline{X} \setminus X) \leq d(x_1, \sigma) \leq d(x_1, x_2) + d(x_2, \sigma).$$

Therefore it follows

$$\operatorname{ir}(x_1) \leq d(x_1, x_2) + \inf_{\sigma \in \overline{X} \setminus X} d(x_2, \sigma) = d(x_1, x_2) + \operatorname{ir}(x_2)$$

and interchanging x_1 and x_2 yields $|\operatorname{ir}(x_1) - \operatorname{ir}(x_2)| \leq d(x_1, x_2)$. So the map is Lipschitz continuous with Lipschitz constant 1. □

The discussion on the immersion radius shows that for well-immersed saddle connections, there exists a lower bound on the length of intersecting saddle connections. We are especially interested in translation surfaces for which there exist enough saddle connections with such a lower bound on the length of intersecting saddle connections. We say that the corresponding singularities fulfill *xossiness* – short for e*x*istence *o*f *s*hort *s*addle connections *i*ntersected *n*ot by *e*ven *s*horter *s*addle connections.

Definition 4.7 (Xossiness)
Let $(X, \mathcal{A})$ be a translation surface and σ a singularity. We say that *σ fulfills xossiness* if for every $\epsilon > 0$ there exists a saddle connection s that connects σ to itself, that has length less than ϵ, and so that there exists a $\delta := \delta(s) > 0$ such that no saddle connection of length less than δ intersects s.

Here and in the following, by saying "two saddle connections do not intersect" we mean that the images of the interiors of the saddle connections do not intersect. If we consider two saddle connections that connect the same singularity to itself then their images naturally have a common point, namely the singularity.

4.4 Conditions for singularities to fulfill xossiness

As the lengths of saddle connections starting in a fixed cone angle or infinite angle singularity are bounded away from 0, cone angle and infinite angle singularities do not fulfill xossiness. For a wild singularity σ, we give two conditions when σ fulfills xossiness. However, the author conjectures that every wild singularity fulfills xossiness.

Proposition 4.8 (Xossiness and the geodesic flow)
Let $(X, \mathcal{A})$ be a translation surface and σ a wild singularity. Suppose that for a dense set of directions, for almost every point, the geodesic flow is defined for all time. Then σ fulfills xossiness.

Proof. Fix $\epsilon' > 0$ so that σ is the only singularity in $B(\sigma, \epsilon') \subseteq \overline{X}$ and choose $\epsilon > 0$ with $\epsilon < \epsilon'$. By Proposition 4.4, there exists a saddle connection $s\colon [0, l] \to \overline{X}$ of length l less than $\frac{\epsilon}{2}$.

If there exists no saddle connection of length less than $\frac{\epsilon}{2}$ that intersects s then s is a saddle connection as desired.

Now suppose there exists a saddle connection $s'\colon [0, l'] \to \overline{X}$ of length l' less than $\frac{\epsilon}{2}$ that intersects s in $s(t)$ with $t \in (0, l)$. Choose a direction θ so that for almost every point the geodesic flow F_θ is defined for all time and so that θ is so close to the direction of s' that $F_\theta(s'(0), [0, l'])$ intersects s

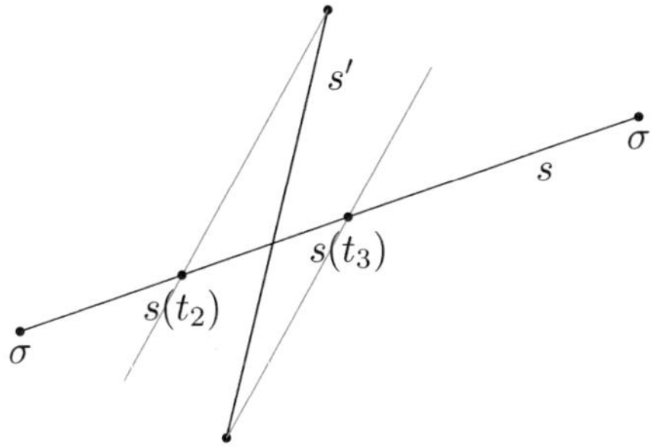

Figure 4.8: A saddle connection s from σ to σ that is intersected by a saddle connection s' and two trajectories of the geodesic flow in a direction close to the direction of s'.

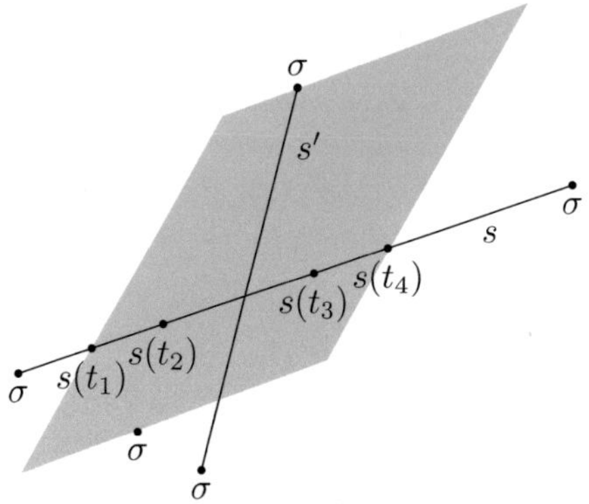

Figure 4.9: The segment $s([t_1, t_4)]$ is flowed forward and backward by F_θ until it hits a singularity for the first time.

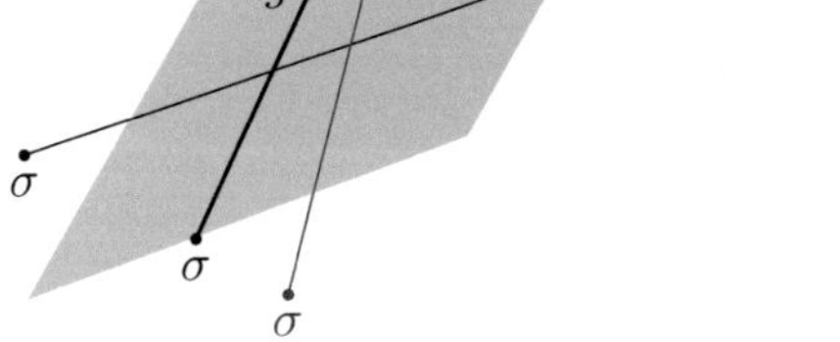

Figure 4.10: The new saddle connection g can not be intersected by saddle connections that are shorter than the distance of g to the left edge and to the right edge of the parallelogram.

in $s(t_2)$ with $t_2 \in (0,t)$ and $F_\theta(s'(l'),[0,-l'])$ intersects s in $s(t_3)$ with $t_3 \in (t,l)$ (see Figure 4.8). Furthermore, choose $t_1 \in (0,t_2)$ and $t_4 \in (t_3,l)$ so that $s(t_1)$ and $s(t_4)$ are points for which the geodesic flow F_θ is defined for all time, in particular backward and forward.

Let $s([t_1,t_4])$ flow backward and forward under F_θ until it hits a singularity. This singularity is in both cases σ because of the choice of $\epsilon < \epsilon'$. By this, we obtain an open parallelogram in X that contains a singularity on two opposite edges but no singularities in the vertices as the vertices are images of t_1 or t_4 under F_θ (see Figure 4.9). The height of the parallelogram is at most the length l' of s'.

The geodesic g in the parallelogram between the two appearances of σ on the boundary as in Figure 4.10 is a saddle connection of length at most $l + l' \leq \epsilon$. The singularity σ has a distinct distance to the left edge and to the right edge of the parallelogram and as the interior of the parallelogram is locally flat, this distance is a lower bound for the immersion radius. Therefore, there exists a $\delta > 0$ so that no saddle connection of length less than δ can intersect g. Hence, g is a saddle connection as desired. □

The second criterion was previously published as [Ran14, Proposition 21'].

Proposition 4.9 (Xossiness and rotational components)
Let $(X,\mathcal{A})$ be a translation surface and σ a wild singularity. Suppose that for every rotational component of σ of length exactly π there exists an angular sector $((0,\pi),\epsilon,i_\epsilon)$ so that the image of $f_{((0,\pi),\epsilon,i_\epsilon)}$ is contained in this rotational component. Then σ fulfills xossiness.

Proof. Fix $\epsilon' > 0$ so that σ is the only singularity in $B(\sigma,\epsilon') \subseteq \overline{X}$ and choose $\epsilon > 0$ with $\epsilon < \epsilon'$. By Proposition 4.4, there exists a linear approach $[\gamma]$ for which the longest representative γ has length $l < \frac{\epsilon}{2}$.

For every $t \in (0,l)$, the immersion radius of $\gamma(t)$ is greater than 0 but at most $d(\gamma(t),\sigma) \leq t$. So we can define the *immersion radius along* γ as the map

$$\mathrm{ir}_\gamma\colon (0,l) \to (0,l),\ t \mapsto \mathrm{ir}(\gamma(t)).$$

For every time $t \in (0,l)$, there exists a geodesic in $\overline{X}$ of length $\mathrm{ir}_\gamma(t)$ connecting $\gamma(t)$ to a singularity. Since $t + \mathrm{ir}_\gamma(t) < \frac{\epsilon}{2} + \frac{\epsilon}{2} < \epsilon'$, this singularity is again σ.

To prove that σ fulfills xossiness, we show the existence of a time t_0 so that $\mathrm{ir}_\gamma(t_0)$ is realized by two different geodesics in $\overline{X}$. Then we can

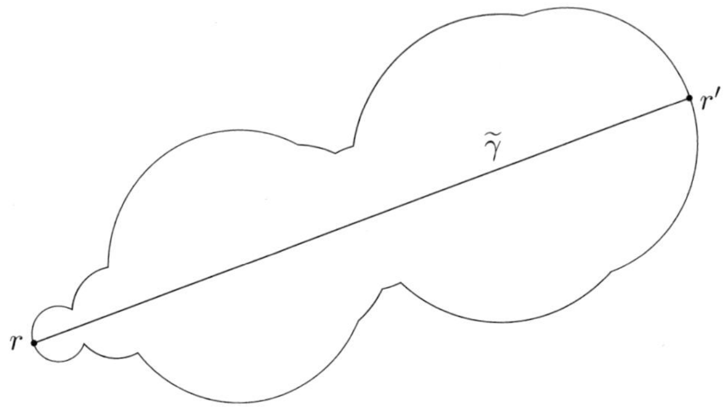

Figure 4.11: The open and simply connected, developed set $\widetilde{B} \subseteq \mathbb{R}^2$.

join the two occurrences of the singularity σ at the end points of the geodesics in $B(\gamma(t_0), \mathrm{ir}_\gamma(t_0))$. The condition on the rotational components of length π makes sure that we obtain a well-immersed saddle connection s for which the immersion radius is defined as in Definition 4.5 (iii). This means that there exists a lower bound on the length of saddle connections that intersect s in its interior.

For every $t \in (0, l)$, there exists a locally flat disk $B(\gamma(t), \mathrm{ir}_\gamma(t))$. We define the locally flat subset

$$B := \bigcup_{t \in (0,l)} B(\gamma(t), \mathrm{ir}_\gamma(t)) \subseteq X.$$

Then we consider a lift $\widetilde{\gamma}$ of γ to the universal cover, together with the union of the disks $B(\widetilde{\gamma}(t), \mathrm{ir}_\gamma(t))$ for every $t \in (0, l)$. By developing this simply connected set into the plane along $\widetilde{\gamma}$, we obtain an open, connected subset $\widetilde{B} \subseteq \mathbb{R}^2$ (see Figure 4.11). The developed curve in $\widetilde{B} \subseteq \mathbb{R}^2$ which corresponds to the curve γ in B is still called $\widetilde{\gamma}$. Every time we encounter σ on the boundary of $B \subseteq X$, we consider a *representative* r of σ on the boundary of $\widetilde{B} \subseteq \mathbb{R}^2$. Note that the singularity σ is always the same in $\partial B \subseteq \overline{X}$ while the representatives are not identified in $\partial \widetilde{B} \subseteq \mathbb{R}^2$.

Define the set $R := \{r \in \partial \widetilde{B} : r \text{ is a representative of } \sigma\} \subseteq \mathbb{R}^2$. Every sequence in R converging to a point $x \in \mathbb{R}^2$ corresponds to a sequence in $\overline{X}$

where all elements are σ, so the limit is σ and hence x is a representative of σ. Therefore, R is a closed set in $\mathbb{R}^2$.

For every representative $r \in R$ we define the set

$$T_r := \{t \in (0,l) : d(\widetilde{\gamma}(t), r) = \mathrm{ir}_\gamma(t)\}.$$

Then the set T_r is closed in $(0,l)$ and connected:

- Let $(t_n)_{n\in\mathbb{N}} \subseteq T_r$ be a sequence converging to a time $t \in (0,l)$. We have

  $$d(\widetilde{\gamma}(t_n), r) \leq d(\widetilde{\gamma}(t), r) + d(\widetilde{\gamma}(t), \widetilde{\gamma}(t_n)) = d(\widetilde{\gamma}(t), r) + |t - t_n|$$

  and

  $$d(\widetilde{\gamma}(t_n), r) \geq d(\widetilde{\gamma}(t), r) - d(\widetilde{\gamma}(t), \widetilde{\gamma}(t_n)) = d(\widetilde{\gamma}(t), r) - |t - t_n|.$$

  As $\mathrm{ir}\colon X \to (0,\infty]$ is continuous (see Lemma 4.6), ir_γ is also continuous and we deduce

  $$\mathrm{ir}_\gamma(t) = \lim_{n\to\infty} \mathrm{ir}_\gamma(t_n) = \lim_{n\to\infty} d(\widetilde{\gamma}(t_n), r) = d(\widetilde{\gamma}(t), r).$$

  So t is in T_r and therefore T_r is closed in $(0,l)$.

- For the connectedness consider t_1, t_3 in T_r and $t_2 \in (0,l)$ such that $t_1 < t_2 < t_3$. Then the circle around $\widetilde{\gamma}(t_2)$ through r is contained in the closure of $B(\widetilde{\gamma}(t_1), \mathrm{ir}_\gamma(t_1)) \cup B(\widetilde{\gamma}(t_3), \mathrm{ir}_\gamma(t_3))$ (see Figure 4.12). This implies that for every $r' \in R$ with $d(\widetilde{\gamma}(t_2), r') < d(\widetilde{\gamma}(t_2), r)$ it also holds $d(\widetilde{\gamma}(t_1), r') < d(\widetilde{\gamma}(t_1), r)$ or $d(\widetilde{\gamma}(t_3), r') < d(\widetilde{\gamma}(t_3), r)$. Because it holds $t_1, t_3 \in T_r$, this is impossible and therefore we have $d(\widetilde{\gamma}(t_2), r) \leq d(\widetilde{\gamma}(t_2), r')$ for every $r' \in R$ and hence $t_2 \in T_r$.

We will now continue with a case-by-case analysis of $\partial\widetilde{B}$ and how it contains R:

Case 1: There is an open, connected subset of $\partial\widetilde{B}$ which is disjoint from R. Then there exists a closed connected subset b of $\partial\widetilde{B}$ whose interior is disjoint from R but whose end points (in a relative sense) are contained in R. We call these end points r_1 and r_2.

For technical reasons, we now consider half-disks instead of disks $B(\gamma(t), \mathrm{ir}_\gamma(t))$ and slightly abuse notation. This means, we only consider the connected components of $B(\widetilde{\gamma}(t), \mathrm{ir}_\gamma(t)) \setminus \mathrm{im}(\widetilde{\gamma})$ which are on

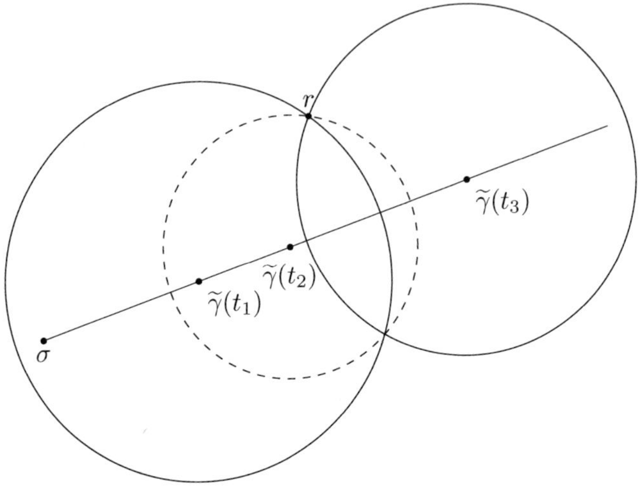

Figure 4.12: The circle around $\widetilde{\gamma}(t_2)$ through r is contained in the closure of the disks around $\widetilde{\gamma}(t_1)$ and $\widetilde{\gamma}(t_3)$ through r.

the same side of γ as r_1 and r_2. Also, we only consider representatives in R on the same side as r_1 and r_2, in T_r we consider times t where the geodesic from $\widetilde{\gamma}(t)$ to r is the shortest on the correct side, $\mathrm{ir}_\gamma(t)$ is the minimum of lengths of geodesics on the correct side, and so on.

In this sense we have that $T_{r_1} \cup T_{r_2}$ is connected: Choose $t_1 \in T_{r_1}$, $t_2 \in T_{r_2}$ and assume $t_1 < t_2$ (if $t_1 > t_2$, change notation). Furthermore, choose $t' \in (0, l)$ with $t_1 < t' < t_2$ (see Figure 4.13). Then we have for every $r \in R \setminus \{r_1, r_2\}$ on the same side of $\widetilde{\gamma}$ as r_1 and r_2 that the inequality $d(\widetilde{\gamma}(t'), r) \geq \min\{d(\widetilde{\gamma}(t'), r_1), d(\widetilde{\gamma}(t'), r_2)\}$ holds, by similar geometric arguments as in the proof of the connectedness of T_r. This means $t' \in T_{r_1} \cup T_{r_2}$ and $T_{r_1} \cup T_{r_2}$ is connected.

As T_{r_1} and T_{r_2} are closed and their union is connected, there exists a point $t_0 \in T_{r_1} \cap T_{r_2}$. In particular, the geodesics from $\widetilde{\gamma}(t_0)$ to r_1 and from $\widetilde{\gamma}(t_0)$ to r_2 in $\widetilde{B} \subseteq \mathbb{R}^2$ have the same length and the corresponding two geodesics from $\gamma(t_0)$ to σ in $B \subseteq X$ are contained in $B(\gamma(t_0), \mathrm{ir}_\gamma(t_0))$ except of their end points. So we can join the end points of the two geodesics in $\overline{B} \subseteq \overline{X}$ and obtain a saddle connection from σ to σ of length less than $2 \cdot \mathrm{ir}_\gamma(t_0) \leq 2 \cdot t_0 \leq 2l < \epsilon$ (see Figure 4.14).

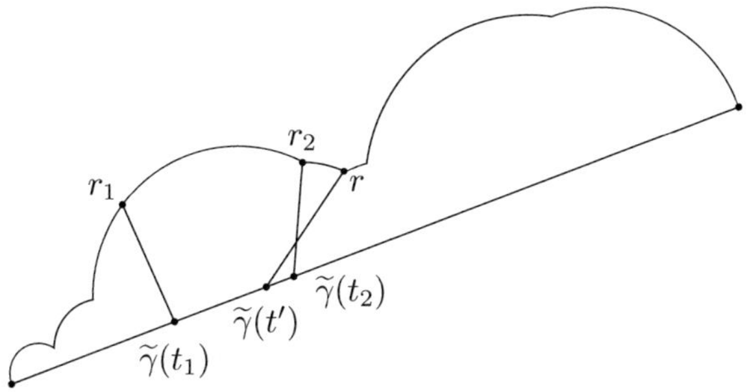

Figure 4.13: If there exists no representative between r_1 and r_2 then t' is contained in $T_{r_1} \cup T_{r_2}$.

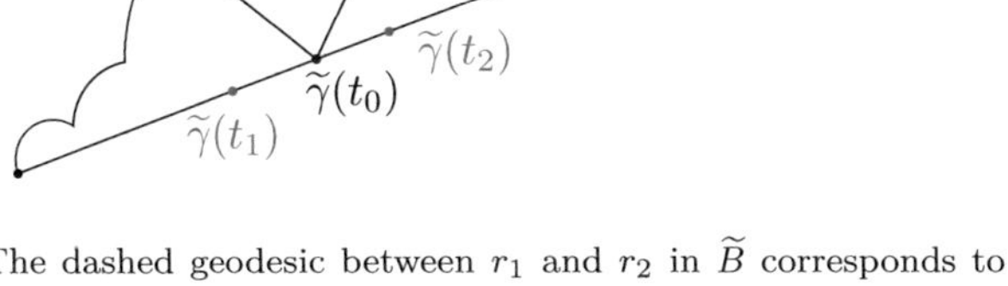

Figure 4.14: The dashed geodesic between r_1 and r_2 in $\widetilde{B}$ corresponds to a saddle connection in B.

The saddle connection from this construction is a chord of the half-disk $B(\gamma(t_0), \mathrm{ir}_\gamma(t_0))$, so the two corresponding linear approaches are contained in rotational components of length at least π. By the remark after Definition 4.5 (for rotational components of length strictly greater than π) and by assumption (for rotational components of length exactly π) there exist angular sectors $((0,\pi), \epsilon_+, i_{\epsilon_+})$ and $((0,\pi), \epsilon_-, i_{\epsilon_-})$ so that the linear approaches defined by the saddle connection are contained in one of the images of $f_{((0,\pi),\epsilon_+,i_{\epsilon_+})}$ and $f_{((0,\pi),\epsilon_-,i_{\epsilon_-})}$. This means that the saddle connection is well-immersed and that there exists a lower bound on the length of saddle connections that intersect the resulting saddle connection in its interior.

Case 2: Every open, connected subset of $\partial\widetilde{B}$ contains a representative r. This means, R is dense in $\partial\widetilde{B}$. As R is a closed set, this implies $R = \partial\widetilde{B}$. When considering the situation in X, we have $\{\sigma\} = \partial B$, hence no point in B can be connected to a point in $X \setminus B$. As X is connected, it follows $X = B$. Consider two parallel geodesics in X, which are close to each other and end in σ, and a Cauchy sequence on each of these geodesics. Then the distance of the corresponding elements of the Cauchy sequences is bounded away from 0. This means that the limits of the Cauchy sequences are two different points but this is a contradiction as we only have one singularity in $\partial B = \{\sigma\}$.

As Case 2 can never happen and in Case 1, for each $\epsilon > 0$ we find a saddle connection as desired, the statement is proven. □

We have now seen in the proof how the peculiar condition on the existence of an angular sector $((0,\pi), \epsilon, i_\epsilon)$ for every rotational component of length exactly π is used. In particular, if such an angular sector does not exist then it is possible in Case 1 that short chords of the disk $B(\gamma(t_0), \mathrm{ir}_\gamma(t_0))$ intersect the obtained saddle connection in its interior (cf. Figure 4.14). We can encounter this behaviour in the following example which is inspired by the parabola surface of Hooper in [Hoo14] and the double parabola surface of Bowman and Valdez in [BV13, Example 1.12].

Example 4.10 (*Hourglass surface*). Consider the two generalized polygons with the vertices $\{(\frac{1}{n}, \frac{1}{n^2}) : n \in \mathbb{Z} \setminus \{0\}\}$ and $\{(\frac{1}{n}, -\frac{1}{n^2}) : n \in \mathbb{Z} \setminus \{0\}\}$ (see Figure 4.15). For each edge there exists a parallel edge of the same length, these pairs of edges are identified to obtain the *hourglass surface*. It has one singularity with four rotational components: Two of the rotational

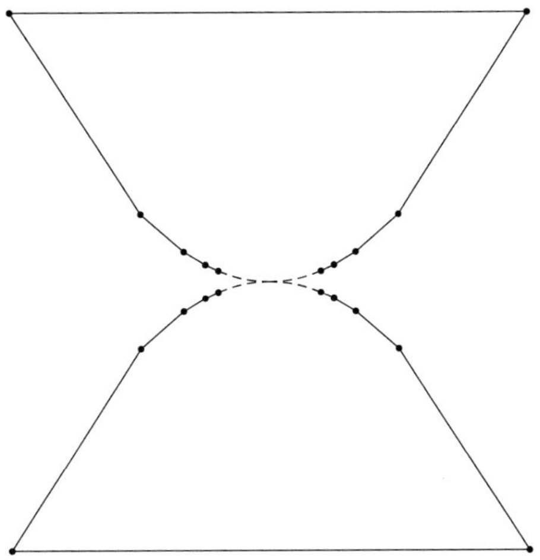

Figure 4.15: For the hourglass surface, we identify parallel edges.

components are of infinite length, the other two are defined by the geodesic curves $\gamma_1 \colon (0,1) \to X, t \mapsto (0,t)$ and $\gamma_2 \colon (0,1) \to X, t \mapsto (0,-t)$ and have length π.

Note that for every $\delta > 0$, there exists an $\epsilon > 0$ and an angular sector $((\delta, \pi - \delta), \epsilon, i_\epsilon)$ with base point $(0,0)$ so that $[\gamma_1] \in \operatorname{im}(f_{((\delta,\pi-\delta),\epsilon,i_\epsilon)})$. However, there does not exist such an angular sector for $\delta = 0$ due to the short edges of the generalized polygon close to $(0,0)$. In particular, for every saddle connection starting in $(0,0)$ there exist horizontal saddle connections of arbitrarily small length that are intersecting the given saddle connection.

Nevertheless, the hourglass surface fulfills xossiness: A saddle connection defined by an edge of the generalized polygons can not be intersected by a saddle connection that is shorter than the edge. Furthermore, there exist arbitrarily short saddle connections of this kind.

5 The influence of wild singularities on the genus

As we mentioned earlier, most known examples of wild translation surfaces with exactly one singularity are Loch Ness monsters, i.e. have infinite genus and one end. The number of ends can easily be increased by gluing in half-cylinders but it is hard to reduce the genus. This leads to the conjecture that the existence of a wild singularity implies infinite genus.

In full generality, this conjecture is not true as we will see in Example 5.9 but we will provide necessary conditions for a wild singularity to imply infinite genus. The conditions involve that the singularity fulfills xossiness and the proofs are based on the conclusions from Section 4.3. The results in this chapter have been previously published in [Ran14].

5.1 Saddle connections as a tool for detecting infinite genus

Our purpose is to show that a translation surface with certain conditions has infinite genus. Therefore, we give a feasible criterion for the infinity of genus in this section.

Recall from Definition 1.26 that we defined the genus of a surface as half of the maximum cardinality of a nonseparating set of curves in X. In particular, a surface has infinite genus if there exists a nonseparating set of infinitely many curves in X. In Proposition 5.1, we specialize this definition to a criterion for the case of translation surfaces using saddle connections. The criterion involves curves that connect one side of a given curve to the other side. We call these curves *left-to-right curves* which is made precise in Definition 5.2 and Definition 5.3. The main result of this section is summarized in the following proposition which will serve as a tool to determine infinite genus in Section 5.2.

Proposition 5.1 (Saddle connections and infinite genus)
Let $(X, \mathcal{A})$ be a translation surface and σ a singularity. Suppose that for every $n \geq 1$ there exists a set of n saddle connections from σ to itself so that the saddle connections intersect exactly in σ and the set has left-to-right curves. Then X has infinite genus.

We will prove the proposition by several lemmas and start with a precise definition of left-to-right curves.

Definition 5.2 (Left-to-right curves of curves in X)
Let X be a connected, orientable surface, $n \geq 1$, and $\gamma_1, \ldots, \gamma_n$ simple closed curves in X that intersect pairwise in exactly one point $x \in X$.

(i) Let $\epsilon > 0$ be small enough so that the ϵ-neighborhood N of γ_1 is a tubular neighborhood. Then N is topologically an annulus. So $N \setminus \operatorname{im}(\gamma_1)$ consists of two connected components N_l and N_r. Considering the underlying orientation of $\operatorname{im}(\gamma_1)$ we call points in N_l and N_r *points on the left of* γ_1 and *points on the right of* γ_1, respectively.

(ii) A curve in $X \setminus \operatorname{im}(\gamma_1)$ from a point on the left of γ_1 to a point on the right of γ_1 is called *left-to-right curve* of γ_1.

(iii) Choose a tubular neighborhood N of γ_n and let N_l and N_r be as before. We have that $N_l \setminus (\operatorname{im}(\gamma_1) \cup \ldots \cup \operatorname{im}(\gamma_{n-1}))$ consists of one or more connected components (see Figure 5.1). The boundary of such a connected component consists of a subset of the boundary of N and of subsets of the images of some γ_i. As the curves intersect in exactly one point, there is only one connected component N_l^* whose boundary additionally contains $\operatorname{im}(\gamma_n)$.

We call a point in this connected component N_l^* *point on the left of* γ_n *with respect to* $\gamma_1, \ldots, \gamma_{n-1}$. Similarly, we can define *points on the right of* γ_n *with respect to* $\gamma_1, \ldots, \gamma_{n-1}$. Then a curve in $X \setminus (\operatorname{im}(\gamma_1) \cup \ldots \cup \operatorname{im}(\gamma_n))$ is called *left-to-right curve of* γ_n *with respect to* $\gamma_1, \ldots, \gamma_{n-1}$ if it connects a point on the left of γ_n with respect to $\gamma_1, \ldots, \gamma_{n-1}$ to a point on the right of γ_n with respect to $\gamma_1, \ldots, \gamma_{n-1}$.

(iv) We say that the set of curves $\{\gamma_1, \ldots, \gamma_n\}$ *has left-to-right curves* if every curve has a left-to-right curve with respect to the other ones.

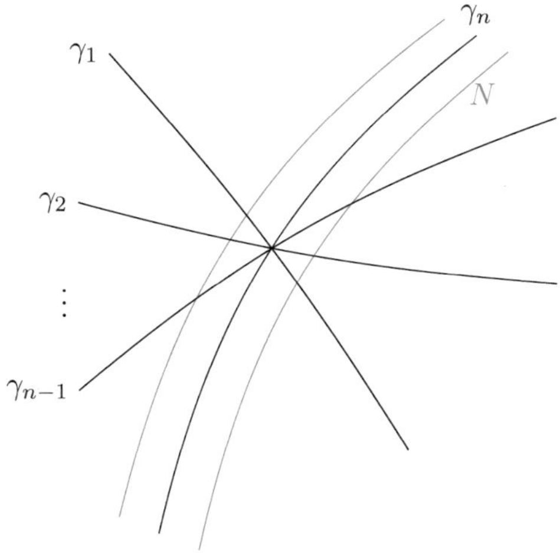

Figure 5.1: The connected components N_l^* and N_r^* are the ones that are only partially drawn in the sketch.

Now let $(X, \mathcal{A})$ be a translation surface. For a curve in $\overline{X}$, for instance a saddle connection, we cannot find such a tubular neighborhood N as in Definition 5.2 (i) but use a slightly different neighborhood instead. So we can define left-to-right curves for a special type of curves in $\overline{X}$ in a similar way while avoiding the singularity.

Definition 5.3 (Left-to-right curves of curves in $\overline{X}$)
Let $(X, \mathcal{A})$ be a translation surface, σ a singularity, $n \geq 1$, and $\gamma_1, \ldots, \gamma_n$ simple closed curves in $X \cup \{\sigma\}$ whose images contain σ exactly as start and end point and that are disjoint in their interiors.

(i) Let l be the length of γ_1 and let $\epsilon > \epsilon' > 0$ be sufficiently small. Consider the set $N \subseteq X \cup \{\sigma\}$ which is the union of $B(\sigma, \epsilon)$ and the open ϵ'-neighborhood $\widetilde{N}$ of the segment $\gamma_1([\epsilon, l - \epsilon])$. Again, $\widetilde{N} \setminus \operatorname{im}(\gamma_1)$ consists of two connected components $\widetilde{N}_l$ and $\widetilde{N}_r$. In this situation, we call points in $\widetilde{N}_l$ *points on the left of* γ_1 and points in $\widetilde{N}_r$ *points on the right of* γ_1, with respect to the orientation of γ_1.

(ii) A curve in $X \setminus \operatorname{im}(\gamma_1)$ from a point on the left of γ_1 to a point on the right of γ_1 is called *left-to-right curve* of γ_1.

(iii) Choose a neighborhood N of γ_n and let $\widetilde{N}_l$ and $\widetilde{N}_r$ be as before. Again, it is possible that $\widetilde{N}_l \setminus (\mathrm{im}(\gamma_1) \cup \ldots \cup \mathrm{im}(\gamma_{n-1}))$ consists of more than one connected component. In this case, we can avoid the indicated behaviour by choosing $\epsilon' > 0$ small enough so that none of the curves $\gamma_1, \ldots, \gamma_{n-1}$ intersects $\widetilde{N}_l$ and $\widetilde{N}_r$. We call a point in these newly chosen $\widetilde{N}_l$ and $\widetilde{N}_r$ *point on the left of γ_n with respect to $\gamma_1, \ldots, \gamma_{n-1}$* and *point on the right of γ_n with respect to $\gamma_1, \ldots, \gamma_{n-1}$*, respectively.

Then a curve in $X \setminus (\mathrm{im}(\gamma_1) \cup \ldots \cup \mathrm{im}(\gamma_n))$ is called *left-to-right curve of γ_n with respect to $\gamma_1, \ldots, \gamma_{n-1}$* if it connects a point on the left of γ_n with respect to $\gamma_1, \ldots, \gamma_{n-1}$ to a point on the right of γ_n with respect to $\gamma_1, \ldots, \gamma_{n-1}$.

(iv) We say that the set of curves $\{\gamma_1, \ldots, \gamma_n\}$ *has left-to-right curves* if every curve has a left-to-right curve with respect to the other ones.

Note that the existence of left-to-right curves as in Definition 5.2 and Definition 5.3 does not depend on the choice of ϵ or ϵ' as long as these values are small.

By means of left-to-right curves we can now formulate a criterion for a set of curves to be nonseparating.

Lemma 5.4 (*Criterion for simple closed curves to be nonseparating*). Let X be a connected, orientable surface, $n \geq 1$, and $\gamma_1, \ldots, \gamma_n$ simple closed curves in X that intersect pairwise in exactly one point $x \in X$. Then the set $\{\gamma_1, \ldots, \gamma_n\}$ is nonseparating if and only if the set has left-to-right curves.

Proof. We prove this statement by induction on the number n of curves. For the base case we show that the curve γ_1 is nonseparating if and only if it has a left-to-right curve.

If γ_1 is nonseparating then we can take any point on the left and any point on the right of γ_1 and as $X \setminus \mathrm{im}(\gamma_1)$ is connected there exists a curve connecting these two points without intersecting γ_1.

Now assume we have such a left-to-right curve δ_1 connecting a point x_l on the left of γ_1 and a point x_r on the right of γ_1. Choose two points $x_1, x_2 \in X \setminus \mathrm{im}(\gamma_1)$. We have to show that there exists a curve β in $X \setminus \mathrm{im}(\gamma_1)$ that connects x_1 and x_2.

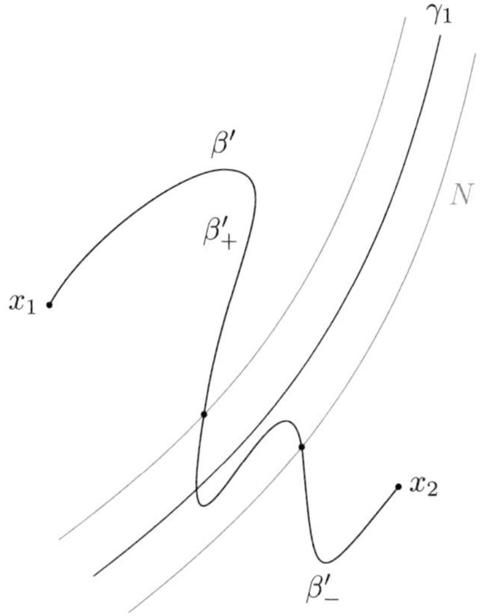

Figure 5.2: This configuration of γ_1 and β' is treated in Case 2.

Let N, N_l and N_r be as in Definition 5.2. As X is connected, there exists a curve β' in X that connects x_1 to x_2. If β' is disjoint from N then we can choose $\beta := \beta'$. If β' is contained in N then x_1 and x_2 can both be connected to x_l or to x_r by a curve that does not intersect γ_1. In particular, concatenating these curves with δ_1 in the right order, x_1 and x_2 can be connected by a curve that does not intersect γ_1.

In all other cases, let β'_+ be the subcurve of β' from x_1 to the first intersection of β' and ∂N and let β'_- be the subcurve of β' from the last intersection of β' and ∂N to x_2 (see Figure 5.2 for a sketch).

Case 1: The end point of β'_+ and the start point of β'_- belong both to ∂N_r or both to ∂N_l. Then we can choose a curve between these two points in the connected set N_r or N_l and the concatenation of this curve with β'_+ and β'_- gives us a curve β as desired.

Case 2: The end point of β'_+ belongs to ∂N_l or to ∂N_r and the start point of β'_- belongs to the other. Without loss of generality, let β'_+ end in ∂N_l and β'_- start in ∂N_r. Again, as N_l and N_r are connected we find curves connecting the end point of β'_+ to x_l in N_l and connecting x_r to the start point of β'_- in N_r. Concatenating all these curves with the left-to-right curve δ_1 in the correct order yields a curve β as desired.

This concludes the proof of the base case. For the inductive step, let $n \geq 2$ and $\{\gamma_1, \ldots, \gamma_n\}$ be a set of simple closed curves that intersect pairwise exactly in x and so that the set $\{\gamma_1, \ldots, \gamma_{n-1}\}$ is nonseparating if and only if the set has left-to-right curves.

Let $\{\gamma_1, \ldots, \gamma_n\}$ be a nonseparating set of curves. We show for every $i \in \{1, \ldots, n\}$ that γ_i has a left-to-right curve by choosing a point on the left and a point on the right of γ_i with respect to $\gamma_1, \ldots, \gamma_{i-1}, \gamma_{i+1}, \ldots, \gamma_n$. As $X \setminus \{\mathrm{im}(\gamma_1) \cup \ldots \cup \mathrm{im}(\gamma_n)\}$ is connected there exists a curve connecting the two chosen points without intersecting one of the curves $\gamma_1, \ldots, \gamma_n$.

Now assume that the set $\{\gamma_1, \ldots, \gamma_n\}$ has left-to-right curves. Then $\{\gamma_1, \ldots, \gamma_{n-1}\}$ has left-to-right curves and is hence nonseparating. Let δ_n be the left-to-right curve of γ_n with respect to $\gamma_1, \ldots, \gamma_{n-1}$ connecting a point x_l on the left of γ_n to a point x_r on the right of γ_n.

As in the base case, we have to show again that for a given choice of two points $x_1, x_2 \in X \setminus (\mathrm{im}(\gamma_1) \cup \ldots \cup \mathrm{im}(\gamma_n))$ there exists a curve β_n in $X \setminus (\mathrm{im}(\gamma_1) \cup \ldots \cup \mathrm{im}(\gamma_n))$ that connects x_1 and x_2. As $\{\gamma_1, \ldots, \gamma_{n-1}\}$ is nonseparating, there exists a curve β'_n in $X \setminus (\mathrm{im}(\gamma_1) \cup \ldots \cup \mathrm{im}(\gamma_{n-1}))$ that connects x_1 and x_2.

If β'_n does not intersect γ_n then we can choose $\beta_n := \beta'_n$. If it does, it also intersects one of the connected components N_l^* or N_r^* of $N \setminus (\mathrm{im}(\gamma_1) \cup \ldots \cup \mathrm{im}(\gamma_n))$. This is because the boundaries of all other connected components contain only one point of $\mathrm{im}(\gamma_n)$ which is the point $x = \mathrm{im}(\gamma_1) \cap \ldots \cap \mathrm{im}(\gamma_n)$. Now let β'_+ be the subcurve of β'_n from x_1 to the first intersection of β'_n and $\partial(N_l^* \cup N_r^*)$ and let β'_- be the subcurve of β'_n from the last intersection of β'_n and $\partial(N_l^* \cup N_r^*)$ to x_2. Now we proceed as in the base case and construct a curve β_n in $X \setminus (\mathrm{im}(\gamma_1) \cup \ldots \cup \mathrm{im}(\gamma_n))$ that connects x_1 and x_2.

This concludes the proof of the inductive step and hence the proof of the lemma. $\square$

It follows by the same arguments that the criterion in Lemma 5.4 is also true for saddle-connections or, more generally, simple closed curves in $\overline{X}$ whose image contains singularities exactly as start and end point. While the beginning of the proof of the base case is literally the same, we have to choose an $\epsilon > 0$ small enough so that all intersection points of β' and γ_1 are contained in the ϵ'-neighborhood $\widetilde{N}$ of the segment $\gamma_1([\epsilon, l-\epsilon])$. Then we can use $\widetilde{N}$ instead of N and finish the proof of the base case in the same way as before. The replacement of N by $\widetilde{N}$ also makes the proof of the inductive step work for this type of curves as the components of $\widetilde{N} \setminus \mathrm{im}(\gamma_1)$ can also play the role of the connected components N_l^* and N_r^*.

Altogether, we have the following version of Lemma 5.4.

Lemma 5.5 (*Criterion for curves in $\overline{X}$ to be nonseparating*). Let $(X, \mathcal{A})$ be a translation surface, σ a singularity, $n \geq 1$, and $\gamma_1, \ldots, \gamma_n$ simple closed curves in $X \cup \{\sigma\}$ whose images contain σ exactly as start and end point and that are disjoint in their interiors. Then the set $\{\gamma_1, \ldots, \gamma_n\}$ is nonseparating if and only if the set has left-to-right curves.

However, as "genus" is a concept for surfaces we have to consider curves in X instead of curves in $\overline{X}$ to determine infinite genus. Because of this, we show in the next lemma how to replace curves in $\overline{X}$ by curves in X without disturbing the left-to-right curves.

Lemma 5.6 (*Nonseparating curves in $\overline{X}$ give rise to nonseparating curves in X*). Let $(X, \mathcal{A})$ be a translation surface, σ a singularity, $n \geq 1$, and $\{\gamma_1, \ldots, \gamma_n\}$ a nonseparating set of simple closed curves in $X \cup \{\sigma\}$ whose images contain σ exactly as start and end point and that are disjoint in their interiors. Then there also exists a set of simple closed curves $\{\gamma_1', \ldots, \gamma_n'\}$ in X that is nonseparating.

Proof. By assumption and Lemma 5.5, there exist left-to-right curves δ_i of γ_i with respect to $\gamma_1, \ldots, \gamma_{i-1}, \gamma_{i+1}, \ldots, \gamma_n$ for every $i \in \{1, \ldots, n\}$. Choose $\epsilon > 0$ small enough so that the ϵ-neighborhood $B(\sigma, \epsilon)$ of σ avoids all δ_i and so that $\partial B(\sigma, \epsilon)$ intersects $\mathrm{im}(\gamma_i)$ at least two times for every $i \in \{1, \ldots, n\}$. For every $i \in \{1, \ldots, n\}$, the first intersection point of $\partial B(\sigma, \epsilon)$ and $\mathrm{im}(\gamma_i)$ (with respect to the orientation of $\mathrm{im}(\gamma_i)$) is called x_i^+ and the last intersection point is called x_i^-.

We will now replace the curves $\gamma_1, \ldots, \gamma_n$ by curves in X that have similar properties. For this choose a point $x \in B(\sigma, \epsilon) \setminus \{\sigma\}$ that will play the role of the current intersection point σ. Because of the path-connectedness of $B(\sigma, \epsilon) \cap X$ (see Proposition 4.1) we have a curve in $B(\sigma, \epsilon) \setminus \{\sigma\}$ from x to x_1^+ and a curve from x_1^- to x (see Figure 5.3). Now let γ_1' be the closed curve that is the concatenation of the curve from x

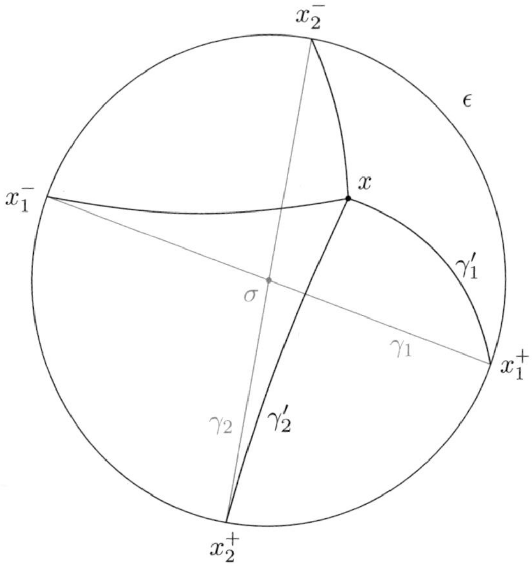

Figure 5.3: The subcurves of the γ_i that are contained in $B(\sigma, \epsilon)$ are replaced by curves in $B(\sigma, \epsilon) \setminus \{\sigma\}$.

to x_1^+, the subcurve of γ_1 from x_1^+ to x_1^-, and the curve from x_1^- to x. If γ_1' intersects itself then we smooth the crossing by joining other pairs of subcurves at the crossing. So without loss of generality γ_1' is simple. Also, as ϵ is chosen small enough, the curve δ_1 is still a left-to-right curve of γ_1' with respect to $\gamma_2, \dots, \gamma_n$.

Now we do the same construction for the rest of the curves successively: For the construction of γ_i' we need a curve from x_i^+ to the prospective intersection point x that does not leave $B(\sigma, \epsilon)$ and does not intersect the curves $\gamma_1', \dots, \gamma_{i-1}'$. We find such a curve by taking any curve from x_i^+ to x in $B(\sigma, \epsilon) \setminus \{\sigma\}$ and instead of possibly crossing some γ_j' we follow γ_j' in a sufficiently small tubular neighborhood without intersecting it until we reach x. Then we define γ_i' as the closed curve that consists of a curve from x to x_i^+ as described, the subcurve of γ_i from x_i^+ to x_i^-, and similarly a curve from x_i^- to x. Again, the curve δ_i is still a left-to-right curve of γ_i' with respect to $\gamma_1', \dots, \gamma_{i-1}', \gamma_{i+1}, \dots, \gamma_n$.

So we have a set of simple closed curves $\gamma_1', \ldots, \gamma_n'$ that are intersecting exactly in x. Also, the set has left-to-right curves $\delta_1, \ldots, \delta_n$ so it is nonseparating by Lemma 5.4. □

By the criterion in Lemma 5.4, we can show that a set of curves is nonseparating but so far we do not have candidates of nonseparating curves for which we could use the criterion. Therefore, we introduce a generalization of the well-known fact that saddle connections of finite translation surfaces are nonseparating.

Lemma 5.7 (Saddle connections are nonseparating). Let $(X, \mathcal{A})$ be a translation surface so that for two directions θ_1, $\theta_2 \in S^1$ the geodesic flows F_{θ_1} and F_{θ_2} are recurrent. Furthermore, let γ be a saddle connection starting and ending at the same singularity. Then γ is nonseparating.

Proof. Consider a geodesic segment $s \subseteq \operatorname{im}(\gamma)$ and the geodesic flow F_θ in a direction $\theta \in S^1$ that is transversal to the direction of γ and so that the flow is recurrent. So there exists a point $x \in s$ that returns to s under the flow F_θ after time t_1. In particular, there exists a time t_0 with $0 < t_0 \leq t_1$ such that $F_\theta(x, t_0) \in \operatorname{im}(\gamma)$ for the first time. Additionally, as γ is geodesic and F_θ is a geodesic flow, the curve $\delta\colon [0, t_0] \to X,\ t \mapsto F_\theta(x, t)$ is approaching $\operatorname{im}(\gamma)$ from the other side than it is leaving. Then for every $\epsilon > 0$ the curve $\delta_\epsilon\colon [\epsilon, t_0 - \epsilon] \to X,\ t \mapsto F_\theta(x, t)$ is a curve in $X \setminus \operatorname{im}(\gamma)$. The curve δ_ϵ or its reversed curve connects a point on the left side of γ to a point on the right side of γ, so it is a left-to-right curve of γ. Hence, γ is nonseparating. □

There exist translation surfaces of infinite but also of finite area so that the geodesic flow is not recurrent for all directions or for all but one direction. We describe such a translation surface in Example 5.9 and note that it has in fact separating saddle connections.

We now have all ingredients to prove Proposition 5.1 which was stated at the beginning of the section.

Proof of Proposition 5.1. As saddle connections are curves that fulfill the conditions of Lemma 5.6, we have a nonseparating set of n curves in X for every $n \geq 1$. By Definition 1.26, this means that the genus of X exceeds every number $n \in \mathbb{N}$ so X has infinite genus. □

5.2 Conditions for wild singularities to imply infinite genus

In this section, we show that the existence of a wild singularity implies infinite genus under certain conditions by using Proposition 5.1. Furthermore, we discuss whether the conditions are necessary and present an example of a translation surface that has a wild singularity and genus 0. We start with the infinite genus result.

Theorem 4 (Wild singularity implies infinite genus)
Let $(X, \mathcal{A})$ be a translation surface so that for two directions $\theta_1, \theta_2 \in S^1$ the geodesic flows F_{θ_1} and F_{θ_2} are recurrent. Furthermore, let σ be a wild singularity of $(X, \mathcal{A})$ that fulfills xossiness. Then X has infinite genus.

Proof. According to the criterion in Proposition 5.1, to prove the statement we have to show for every $n \geq 1$ that there exist n saddle connections from σ to itself that intersect exactly in σ and so that the set has left-to-right curves for every $n \geq 1$. We do this by induction on n.

For the base case of $n = 1$ we choose a saddle connection γ_1 so that there exists a lower bound $\epsilon_1 > 0$ on the length of intersecting saddle connections. Such a saddle connection exists by the assumption that σ fulfills xossiness and by Lemma 5.7 it is nonseparating.

For the inductive step assume that we have a nonseparating set of saddle connections $\{\gamma_1, \ldots, \gamma_{n-1}\}$ for which there exist lower bounds $\epsilon_1, \ldots, \epsilon_{n-1}$ on the lengths of intersecting saddle connections. In particular, for every $i \in \{1, \ldots, n-1\}$ there exists a left-to-right curve δ_i, i.e. a curve in X that connects the left side and the right side of γ_i without intersecting any of the γ_j, $j \in \{1, \ldots, n-1\}$. Let ϵ be the minimum of $\epsilon_1, \ldots, \epsilon_{n-1}$ and of the immersion radii of $\delta_1, \ldots, \delta_{n-1}$. As proven in Proposition 4.4 there exists a saddle connection γ_n from σ to itself with length less than ϵ. Therefore, γ_n does not intersect any of the curves $\gamma_1, \ldots, \gamma_{n-1}, \delta_1, \ldots, \delta_{n-1}$. Additionally, as σ fulfills xossiness we can choose γ_n so that there exists a lower bound $\epsilon_n > 0$ on the length of intersecting saddle connections.

As γ_n is nonseparating in X (see again Lemma 5.7) there exists a left-to-right curve δ'_n in $X \setminus \operatorname{im}(\gamma_n)$. If δ'_n does not intersect $\gamma_1, \ldots, \gamma_{n-1}$ then we can define $\delta_n := \delta'_n$ and δ_n connects the left side of γ_n to the right side of γ_n in $X \setminus (\operatorname{im}(\gamma_1) \cup \ldots \cup \operatorname{im}(\gamma_n))$. Furthermore, none of the curves

$\delta_1, \ldots, \delta_{n-1}$ intersects γ_n. Therefore we have that $X \backslash (\mathrm{im}(\gamma_1) \cup \ldots \cup \mathrm{im}(\gamma_n))$ is connected and the set of curves $\{\gamma_1, \ldots, \gamma_n\}$ has left-to-right curves.

If δ'_n intersects at least one of the curves $\gamma_1, \ldots, \gamma_{n-1}$ then we modify it in the following way. For every intersection with a curve γ_i (without loss of generality, from the left of γ_i) we choose a point x_l on the left and a point x_r on the right of γ_i in $\mathrm{im}(\delta'_n)$. Then we can replace the subcurve of δ'_n that intersects γ_i by a curve in N_l^* from x_l to the start point of δ_i concatenated with δ_i and concatenated with a curve in N_r^* from the end point of δ_i to x_r. By the induction hypothesis and by the choice of ϵ, every δ_i for $i \in \{1, \ldots, n-1\}$ does not intersect any of the curves $\gamma_1, \ldots, \gamma_n$. Therefore the new curve δ_n is a left-to-right curve of γ_n with respect to $\gamma_1, \ldots, \gamma_{n-1}$.

We have thus shown that for every $n \geq 1$, there exists a set of saddle connections of cardinality n which has left-to-right curves. This implies by Proposition 5.1 that X has infinite genus. □

One of the key points of the proof of Theorem 4 is the assumption of recurrence of the geodesic flow. By Poincaré recurrence (see Proposition 1.41) one can deduce recurrence of the geodesic flow from the two weaker conditions that the flow is defined for a set of points of full measure for all time and that the area is finite. We show that none of the two conditions on its own works for the proof by considering the following two examples.

Example 5.8 (*Icicled surface*). Consider a half-open rectangle of height 2 and width 1. The left side is glued to the right side, the bottom and the top are excluded.

For every $n \geq 1$, we consider a vertical segment starting at the bottom and a vertical segment starting at the top, at $\frac{i}{2^n}$ of length $\frac{1}{2^n}$ for every odd $i \in \{1, \ldots, 2^n - 1\}$ (see Figure 5.4). We call the vertical segments *icicles*.

Then we glue the segments as sketched in Figure 5.5. Note that no icicle on the top is glued to an icicle on the bottom. Formally, we can describe the gluings in the following way (starting with the icicles on the top).

- For each side of the icicle at $\frac{1}{2}$, we cut the segment again: first we cut it in half, then we cut the upper half into halves again, cut the upper quarter into halves again, ... So for every $n > 1$ we have a segment of length $\frac{1}{2^n}$.

Figure 5.4: Vertical segments in the icicled surface.

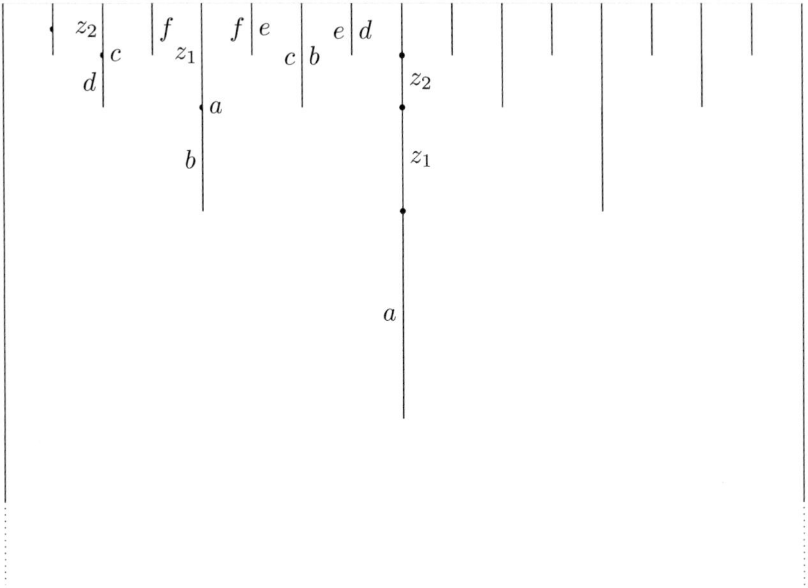

Figure 5.5: Gluings for the icicled surface: segments with the same letters are glued.

- The left side of the lower half of the icicle at $\frac{1}{2}$ is glued to the right side of the icicle at $\frac{1}{4}$.
- For every $n > 2$ and every odd $i \in \{3, \ldots, 2^{n-1} - 1\}$, the left side of the icicle at $\frac{i}{2^n}$ is glued to the right side of the icicle at $\frac{i-2}{2^n}$.
- For every $n > 2$, the left side of the icicle at $\frac{1}{2^{n-1}}$ is cut into two segments of the same length. The lower part is glued to the right side of the icicle at $\frac{2^{n-1}-1}{2^n}$. The upper part is glued to the right side of the segment at $\frac{1}{2}$ which has the correct length.
- We do the similar gluing for the right part of the top and also for the bottom.

The resulting translation surface $(X, \mathcal{A})$ is called *icicled surface* and has the following properties.

(i) There exist exactly two singularities: all the tips of the icicles on the top are identified by the definition of the gluings. We call the corresponding singularity σ_{top}. Now consider a nondyadic point p in the top boundary of the rectangle, i.e. a point where no icicle starts. There exists a sequence of icicles such that the tips of the icicles converge to p, seen as points in $\mathbb{R}^2$ without gluings. Therefore, the distance from σ_{top} to p is 0 and $p = \sigma_{top}$ in $\overline{X}$. The same argument works for the dyadic points on the boundary where an icicle starts. Note that the points where we cut some icicles into more segments are glued to points on the boundary, so these are also equal to σ_{top}.

Also, the same reasoning holds for the tips of the icicles on the bottom and the points in the bottom boundary. So we have only two singularities σ_{top} and σ_{bottom}. In particular, the set of singularities is discrete.

It is worth noting that both of the singularities have one rotational component of infinite length and uncountably many rotational components of finite length.

(ii) For both singularities, every icicle defines a saddle connection or a chain of saddle connections. This means that there exist arbitrarily short saddle connections as it is true for wild translation surfaces in general. However, for all of these saddle connections defined by icicles, there also exist arbitrarily short saddle connections close to the top or bottom, intersecting them. On the other hand, we can define saddle connections from the tip of the icicle at $\frac{1}{2^n}$ to the tip of the icicle at $\frac{1}{2^{n+1}}$ for every $n \geq 1$ (see Figure 5.6). The length of the nth such saddle connection is $\frac{\sqrt{2}}{2^{n+1}}$ and no saddle connection of length smaller than $\frac{\sqrt{2}}{2^{n+1}}$ can intersect it. This means that both singularities fulfill xossiness.

(iii) For a regular point in X, the trajectory under a geodesic flow is only defined until it hits the tip of an icicle or the top or bottom of the rectangle. Hence, there exists no point in X so that the geodesic flow in the vertical direction is defined for all time.

Let θ be a direction in $(0, \pi)$. Suppose that the geodesic flow F_θ is defined for all time for a set of full measure. Consider a closed horizontal geodesic g in the middle of the surface and a tubular

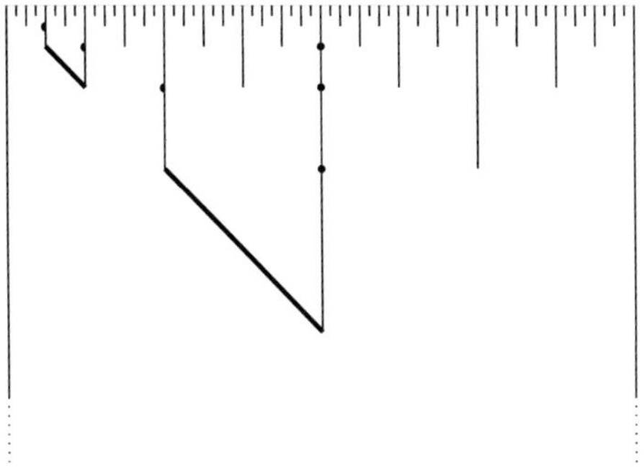

Figure 5.6: The first and the third saddle connection of the set described in Example 5.8 (ii) to show xossiness.

neighborhood N of g not intersecting any icicles. Then F_θ is defined on a subset of N of full measure for all time. From Poincaré recurrence (see Proposition 1.41) we can deduce that there exists a point $x \in N$ and a time $t_x > 0$ so that $F_\theta(x, t_x) \in N$. This means that $F_\theta(x, t_x - \epsilon)$ is contained in the lower part of the surface for an $\epsilon > 0$. But this is impossible as there is no possibility to reach the lower part of the surface from the upper part of the surface.

Hence for all directions except for the horizontal one, there is no set of points of full measure for which the geodesic flow is defined for all time.

(iv) The horizontal flow is periodic but the length of the period is getting the greater the nearer to the top or to the bottom the start point is.

(v) Because all but the horizontal geodesic flow are not recurrent, we cannot use the arguments of Theorem 4 to show infinite genus. On the other hand, we can check by a sharp look that every icicle is defining at least one which has a left-to-right curve. Moreover, every set of saddle connections defined by icicles has left-to-right curves and the number of icicles is not bounded. Therefore, for every $n \geq 1$ there exists a set of n saddle connections that has left-to-right curves. With this we can show directly by Proposition 5.1 instead of Theorem 4 that X has infinite genus.

We want to emphasize that in this example, the top and bottom boundary is in some sense "trapping" the whole geodesic flow. This kind of dynamical behaviour was so far only known for translation surfaces with a continuum of singularities, for instance for the open disk (see the remark at the beginning of Section 1.2). However, we explicitly excluded this kind of examples by requiring that our translation surfaces shall have discrete singularities. There was the expectation that discreteness of the set of singularities should imply good dynamical properties. However, the icicled surface in Example 5.8 shows that there exist translation surfaces with a discrete set of singularities so that for at most one direction θ the geodesic flow F_θ is defined for almost every point for all time. In particular, it is not possible to apply Poincaré recurrence (see Proposition 1.41) to conclude from the finiteness of the area that a geodesic flow is recurrent.

The recurrence of the flow is needed for the proof of Theorem 4 as it is an assumption in Lemma 5.7 which is a crucial ingredient in the proof. In fact, there are translation surfaces (with discrete singularities) that have separating saddle connections. For example, the horizontal saddle connection of the icicled surface that connects the tip of the longest icicle to itself is separating. However, as was indicated in Example 5.8 (v) there exist other saddle connections that are nonseparating. This is not necessarily the case as we will see in the next example which was worked out together with Pat Hooper.

Example 5.9 (Nested cylinders). Consider a Euclidean half-plane with a distinguished midline. We cut vertical slits of infinite length in the half-plane from the midline upward, starting from

$$\frac{1}{2} + \sum_{i=2}^{n} \frac{1}{2i-1} + \frac{1}{2i} \qquad \text{for } n \geq 1.$$

Additionally, we cut vertical slits of infinite length from the midline downward, starting from

$$\sum_{i=1}^{n} \frac{1}{2i} + \frac{1}{2i+1} \qquad \text{for } n \geq 1.$$

Now we glue the right side of a slit to the left side of the slit which is next to the right and the left side of the slit to the right side of the slit

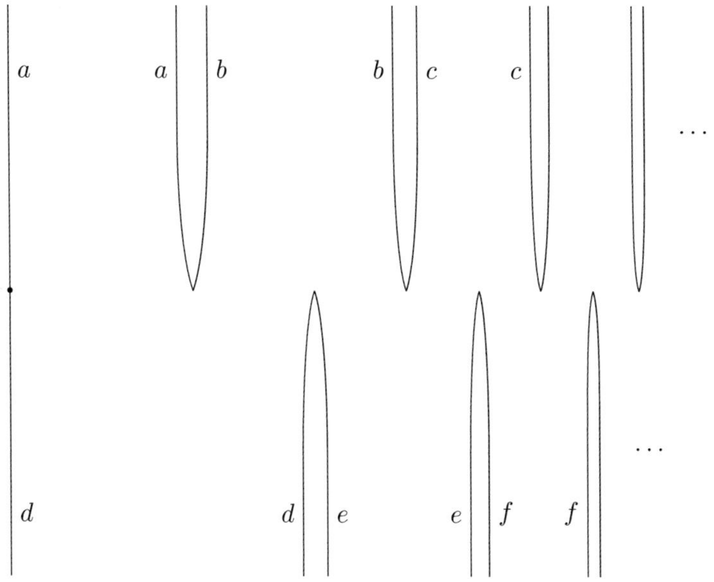

Figure 5.7: The nested cylinders example has a wild singularity and genus 0.

which is next on the left (see Figure 5.7). By this construction, we obtain half-cylinders with smaller and smaller circumferences that are glued in a nested way.

The resulting translation surface has infinite area, genus 0 and exactly one singularity. The singularity is wild as the distance between the start points of the slits is going to 0, i.e. there exists no cyclic translation covering from a punctured neighborhood of σ to a once-punctured disk in $\mathbb{R}^2$. This singularity has exactly one rotational component which is isometric to $\mathbb{R}$.

For every direction θ, the geodesic flow F_θ is defined for almost every point for all time. However, recurrence only occurs in the horizontal direction. In this example, Poincaré recurrence (see Proposition 1.41) is not applicable because the area is not finite.

In particular, all saddle connections are horizontal and all of them are separating.

The last example indicates that the statement in Theorem 4 is wrong if we give up the condition on recurrence in two directions. However, the example does not destroy the prospect of a weaker condition like finite area that could replace the recurrence condition.

Besides, there are a lot of examples of translation surfaces fulfilling the dynamical condition in Theorem 4, i.e. translation surfaces for which there exist two directions so that the geodesic flow in these directions is recurrent. A particular class of such examples are *parabolic translation surfaces*, i.e. translation surfaces that have no Green's function (cf. [Tre14, Remark 1]). For parabolic translation surfaces it follows from [Str84, Theorems 13.1 and 24.4] that for all directions the geodesic flow is recurrent. This implies furthermore, together with Proposition 4.8, that every wild singularity of a parabolic translation surface fulfills xossiness. Therefore, it can be stated that for every wild, parabolic translation surface $(X, \mathcal{A})$ the genus of X is infinite.

6 Spaces of rotational components

The investigation of the space of rotational components is a very recent aspect of the theory of translation surfaces. To use the space of rotational components in a prospective classification of translation surfaces it is important to understand the object on its own. In this chapter, we provide results on the relations between the space of rotational components, the space of linear approaches, and the translation structures on individual rotational components. In the last section, we give instances of topological spaces that can occur as spaces of rotational components.

The results in this chapter were developed together with Lucien Clavier and Chenxi Wu and were published in [CRW14].

6.1 Topological properties of rotational components

In this section, we describe relations between three topologies which are associated to a rotational component of a singularity σ that contains more than one linear approach. These are

(i) the topology on the rotational component from the one-dimensional translation structure on the rotational component, as defined in Definition 1.19,

(ii) the topology of the subspace of $\mathcal{L}(\sigma)$ that contains all linear approaches that are contained in the rotational component, as defined in Definition 1.20, and

(iii) the topology of $\widetilde{\mathcal{L}}(\sigma)$ as a quotient space of $\mathcal{L}(\sigma)$ which contains the rotational component as a point, as defined in Definition 1.24.

We start with a relation between the first and the second topology and will resume with two relations between the first and the third topology.

Lemma 6.1 (*Topology from translation structure refines subspace topology from* $\mathcal{L}(\sigma)$). Let $(X, \mathcal{A})$ be a translation surface, σ a singularity, and c a rotational component of σ that contains more than one linear approach. Furthermore, let $\mathcal{T}_1$ be the topology on c that is induced by the one-dimensional translation structure and let $\mathcal{T}_2$ be the topology on c that is the subspace topology from $\mathcal{L}(\sigma)$.

Then $\mathcal{T}_1$ is finer than $\mathcal{T}_2$.

Proof. Let I be the generalized interval that is isometric to c with the one-dimensional translation structure, $[\gamma]$ a linear approach contained in c, and $y \in I$ the point that corresponds to $[\gamma]$.

Recall from Lemma 1.22 that

$$B(\gamma, t, r) := \widetilde{B}(\gamma(t), r)^t \cap \mathcal{L}(\sigma)$$

is the set of all linear approaches to σ with a representative γ' for which $\gamma'(t)$ is contained in the disk around $\gamma(t)$ of radius r. The collection of sets $B(\gamma, t, r)$ for all $t > 0, r > 0$ is a neighborhood subbasis of $[\gamma]$ in $\mathcal{L}(\sigma)$.

We show that for each $B(\gamma, t, r)$, there exists an open subset of I that contains y and whose points correspond to linear approaches that are contained in $B(\gamma, t, r)$.

Let γ be a representative of $[\gamma]$, $t > 0$ so that $\gamma(t)$ is defined, and $r > 0$. Furthermore, let $U' \subseteq I$ be the collection of points in the interval

$$\left(y - \arccos\left(1 - \frac{r^2}{2t^2}\right), y + \arccos\left(1 - \frac{r^2}{2t^2}\right)\right) \cap I$$

that correspond to linear approaches which are also contained in $\rho_t(\mathcal{L}^t(\sigma))$ where $\rho_t \colon \mathcal{L}^t(\sigma) \hookrightarrow \mathcal{L}(\sigma)$ is the embedding from the direct system. Then U' contains an open subinterval U of I and the linear approaches that correspond to points in U form a subset of $B(\gamma, t, r)$ (cf. the proof of Lemma 2.6 for the argument by the law of cosines). $\square$

We now describe relations between topological properties of a rotational component as a point in $\widetilde{\mathcal{L}}(\sigma)$ and the one-dimensional translation structure on this rotational component.

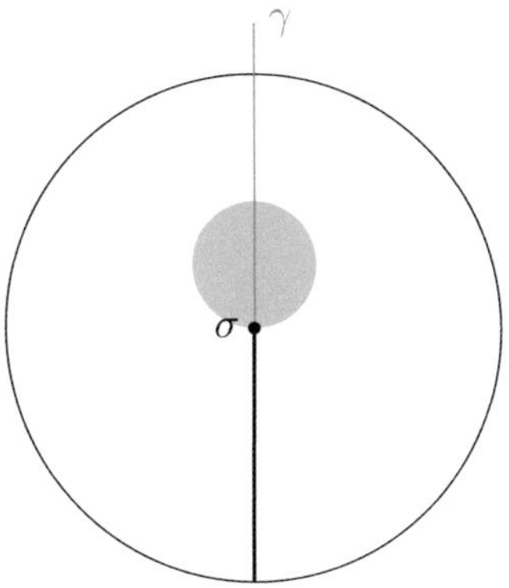

Figure 6.1: The set $i_\epsilon((-\infty, \epsilon) \times (-\pi, \pi))$ as an open disk with a vertical slit and the shaded region $B(\gamma(\frac{\epsilon}{4}), \frac{\epsilon}{4})$.

Proposition 6.2 (Rotational components isometric to $\mathbb{R}$ are open points)
Let $(X, \mathcal{A})$ be a translation surface, σ a singularity, and c a rotational component of σ that contains more than one linear approach. If c with the one-dimensional translation structure is isometric to $\mathbb{R}$ then $\{c\}$ is open in $\widetilde{\mathcal{L}}(\sigma)$.

Proof. For a rotational component $c \in \widetilde{\mathcal{L}}(\sigma)$, the set $\{c\}$ is open in $\widetilde{\mathcal{L}}(\sigma)$ if the set of linear approaches contained in c is open in $\mathcal{L}(\sigma)$. To prove the statement, we find an open neighborhood for every linear approach contained in c that does not contain linear approaches from other rotational components. The union of these neighborhoods is then an open set in $\mathcal{L}(\sigma)$.

Let $c \in \widetilde{\mathcal{L}}(\sigma)$ be a rotational component whose one-dimensional translation structure is isometric to $\mathbb{R}$. Then for every linear approach $[\gamma]$ in c with a representative γ, there exists an angular sector $([-2\pi, 2\pi], \epsilon, i_\epsilon)$ such that $f_{([-2\pi,2\pi],\epsilon,i_\epsilon)}(0) = [\gamma]$. By Lemma 1.22, $B(\gamma, \frac{\epsilon}{4}, \frac{\epsilon}{4})$ is an open neighborhood of $[\gamma]$ in $\mathcal{L}(\sigma)$. We will show that the linear approaches in $B(\gamma, \frac{\epsilon}{4}, \frac{\epsilon}{4})$ are also contained in c.

The set $i_\epsilon((-\infty, c) \times (-\pi, \pi))$ in X is isometric to an open ϵ-disk in $\mathbb{R}^2$ with a slit removed (see Figure 6.1). Let $s\colon (0, \frac{\epsilon}{4}] \to X$ be a geodesic curve with $s(\frac{\epsilon}{4}) \in B(\gamma(\frac{\epsilon}{4}), \frac{\epsilon}{4})$. Then the image of s either intersects the slit or it is contained in $i_\epsilon((-\infty, \epsilon) \times (-\pi, \pi))$.

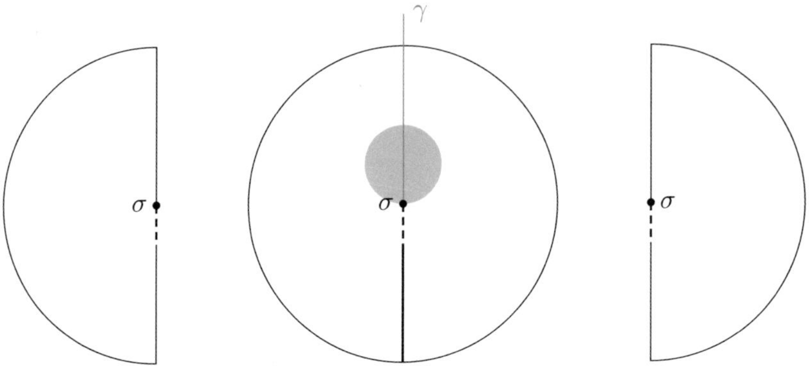

Figure 6.2: The sets $i_\epsilon((-\infty,\epsilon)\times(-2\pi,-\pi))$, $i_\epsilon((-\infty,\epsilon)\times(-\pi,\pi))$, and $i_\epsilon((-\infty,\epsilon)\times(\pi,2\pi))$ (from left to right).

Case 1: If the image of s intersects the slit, it can only do so at the dashed segments $i_\epsilon((-\infty,\frac{\epsilon}{4})\times\{-\pi\})$ or $i_\epsilon((-\infty,\frac{\epsilon}{4})\times\{\pi\})$ in Figure 6.2. Now consider the two sets $i_\epsilon((-\infty,\epsilon)\times(-2\pi,-\pi))$ and $i_\epsilon((-\infty,\epsilon)\times(\pi,2\pi))$, which are open half-disks with radius ϵ as in Figure 6.2. A geodesic segment shorter than $\frac{\epsilon}{4}$ starting on either of the dashed segments, heading inwards, has its endpoint in the interior of this half-disk, hence s can not start in a singularity. This means that s is not contained in $\mathcal{L}^{\frac{\epsilon}{4}}(\sigma)$ and does not define a linear approach to σ.

Case 2: If the image of s is contained in $i_\epsilon((-\infty,\epsilon)\times(-\pi,\pi))$ then it defines a linear approach to σ if it can not be extended to $s(0)$ in X. In this case, s starts in the center of the disk and hence defines a linear approach in the same rotational component c as $[\gamma]$. □

Note that the converse statement is not true in general. For instance, the stack of boxes in Section 2.3 has a singularity σ with exactly one rotational component c. So the set $\{c\}$ is clearly open in $\widetilde{\mathcal{L}}(\sigma)=\{c\}$. On the other hand, the rotational component c in the stack of boxes has a one-dimensional translation structure which is isometric to $(0,\infty)$ and not to $\mathbb{R}$. However, a slightly weaker version of the converse of Proposition 6.2 is true.

Proposition 6.3 (Rotational components as open points have infinite length)
Let $(X, \mathcal{A})$ be a translation surface, σ a wild singularity, and c a rotational component of σ that contains more than one linear approach. If $\{c\}$ is open in $\widetilde{\mathcal{L}}(\sigma)$ then it has a one-dimensional translation structure which is isometric to a connected subset of $\mathbb{R}$ of infinite length.

Proof. Let c be a rotational component of finite length. We show that there exists a sequence of linear approaches not contained in c that converges to a linear approach contained in c. This implies that $\{c\}$ is not open in $\widetilde{\mathcal{L}}(\sigma)$.

As c is of finite length, it is isometric to an interval with end points $a, b \in \mathbb{R}$ (the interval can be open, closed, or none of that). Let $[\gamma]$ be a linear approach which corresponds to a point in the interval that differs strictly less than $\frac{\pi}{2}$ from a.

Choose a representative γ of $[\gamma]$, a regular point $\gamma(t_0)$ on it, and an $\epsilon_1 > 0$ so that $B(\gamma(t_0), \epsilon_1) \subseteq X$ is isometric to a flat disk of radius ϵ_1 (see Figure 6.3). Let

$$t_1 = \inf\{t > 0 : B(\gamma(t), \epsilon_1) \text{ is isometric to a flat disk of radius } \epsilon_1\}.$$

By definition, we have $t_1 \leq t_0$ but we also have $t_1 \geq \epsilon_1 > 0$. Indeed, the disk $B(\gamma(\epsilon_1), \epsilon_1)$ can not be isometric to a flat disk since the point corresponding to $[\gamma]$ in the interval isometric to c would have distance at least $\frac{\pi}{2}$ from a.

By definition of t_1, the boundary of the disk $B(\gamma(t_1), \epsilon_1)$ contains a singularity. As the set of singularities of $(X, \mathcal{A})$ is discrete, for t_0 and ϵ_1 small enough it is the singularity σ which must be contained in the boundary of the disk $B(\gamma(t_1), \epsilon_1)$. This determines a geodesic curve γ_1 in X starting on the boundary of $B(\gamma(t_1), \epsilon_1)$ with the same direction as γ and defined at least on $(0, t_0 - t_1)$ (see Figure 6.3). Therefore the distance between γ and γ_1 in $\mathcal{L}^{t_0 - t_1}(\sigma)$ with the uniform metric is no more than $t_1 + \epsilon_1$.

Repeat this construction with $t_1' \leq t_1$ and $\epsilon_2 > 0$ small enough so that γ_1 is not contained in the $(t_1' + \epsilon_2)$-neighborhood of γ in $\mathcal{L}^{t_0 - t_1}(\sigma)$ anymore. We receive a different linear approach $[\gamma_2]$ closer to $[\gamma]$. By repeating this construction again and again, we iteratively obtain infinitely many different linear approaches $([\gamma_n])_{n \geq 1}$ which all have the same direction.

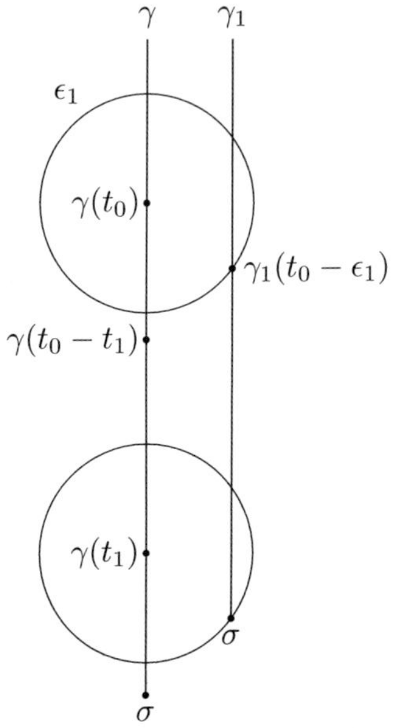

Figure 6.3: For the linear approach $[\gamma]$, there exists a linear approach $[\gamma_1]$ so that the representatives γ and γ_1 are parallel with distance at most ϵ_1.

Only finitely many of them can be contained in the rotational component c as c is of finite length. Thus there is a number $N \geq 1$ so that $[\gamma_n]$ is not contained in c for $n > N$, and $([\gamma_n])_{n>N}$ converges to $[\gamma]$ in c. Therefore $\{c\}$ is not open in $\widetilde{\mathcal{L}}(\sigma)$. □

The proofs of the two preceding lemmas suggest how to tackle the problem of understanding the topology of a rotational component as subspace of $\mathcal{L}(\sigma)$. The interesting linear approaches in terms of this topology are the linear approaches that differ at most π from the boundary of the rotational component with the translation structure. The remaining linear approaches are contained in a subspace of the rotational component which carries the same topology as coming from the translation structure.

We will use this insight in the constructions in the proofs of Theorem 5 and Theorem 6.

6.2 Coarsening from $\mathcal{L}(\sigma)$ to $\widetilde{\mathcal{L}}(\sigma)$

When we pass from the space $\mathcal{L}(\sigma)$ of linear approaches to the quotient space $\widetilde{\mathcal{L}}(\sigma)$ of rotational components, we apparently lose information. However, it is still natural to ask whether we can recover properties of $\mathcal{L}(\sigma)$ by information on $\widetilde{\mathcal{L}}(\sigma)$. In Theorem 5 we show that this is in general not possible as there are uncountably many translation surfaces with non-homeomorphic spaces of linear approaches but with homeomorphic spaces of rotational components.

Theorem 5 ($\widetilde{\mathcal{L}}(\sigma)$ does not determine $\mathcal{L}(\sigma)$)
There are uncountably many translation structures $\mathcal{A}_r$ on a Loch Ness monster of finite area with the following properties.

(i) Every translation surface $(X_r, \mathcal{A}_r)$ has exactly one singularity σ_r.

(ii) The spaces $\widetilde{\mathcal{L}}(\sigma_r)$ of rotational components are all homeomorphic.

(iii) The spaces $\mathcal{L}(\sigma_r)$ of linear approaches are pairwise not homeomorphic.

Proof. Let $\mathbb{K} = \{0,1\}^{\mathbb{N}}$ and consider the equivalence relation on $\mathbb{K}$ defined by $r \sim r'$ if and only if some shifts of r and r' are equal, i.e. if the tails of r and r' agree up to shift.

We will define a translation surface $(X_r, \mathcal{A}_r)$ for every $r \in \mathbb{K}$ but not in the equivalence class of $(0,0,0,\ldots)$ by modifying the stack of boxes in Section 2.3. The translation surface $(X_r, \mathcal{A}_r)$ will have exactly one singularity σ_r and one rotational component. Then we will show that if $\mathcal{L}(\sigma_r)$ and $\mathcal{L}(\sigma_{r'})$ are homeomorphic then r and r' are in the same equivalence class. This will prove the statement since there are uncountably many equivalence classes in $\mathbb{K}$.

Let $r = (r_n) \in \mathbb{K}$ but not in the equivalence class of $(0,0,0,\ldots)$. Recall the stack of boxes from Section 2.3 and choose the two sequences H and W by $h_n = 1$ and $w_n = 2^{-n}$ for every $n \in \mathbb{N}$. We modify the stack of boxes by gluing in additional rectangles, depending on r: Consider vertical geodesic segments v_n of length 2^{-2n} starting at each A_n for $n \geq 2$ (see Figure 6.4). For each $n \geq 2$ such that $r_n = 1$, we cut open along v_n and glue the two sides of the slit v_n to the two vertical sides of a rectangle of width 1, then we identify the top and bottom sides of the rectangle. We call the resulting translation surface $(X_r, \mathcal{A}_r)$. As in Section 2.3, it is topologically a Loch Ness monster and it has finite area. Furthermore, the translation surface has still one singularity σ_r and one rotational component isometric to $(0, \infty)$. Hence the spaces $\widetilde{\mathcal{L}}(\sigma_r)$ are homeomorphic for all r.

We now elaborate tools to recover the equivalence class of r intrinsically from the topology of $\mathcal{L}(\sigma_r)$ for an $r \in \mathbb{K}$. Let c be the rotational component of σ_r with the one-dimensional translation structure. Then there exists an isometry $(0, \infty) \to c$. As c is the only rotational component of σ_r, we also have a bijective map $(0, \infty) \to \mathcal{L}(\sigma_r)$. Let $\mathbb{R}_+^r$ be the set $(0, \infty)$, equipped with the pullback of the topology of $\mathcal{L}(\sigma_r)$. In the following, we use the parameter $x \in \mathbb{R}_+^r$ to represent linear approaches in $\mathcal{L}(\sigma_r)$ and we let x run through $\mathbb{R}_+^r$, i.e. the linear approach is rotating counterclockwise around the singularity.

To record when the rotated linear approach is contained in an additional cylinder, we will use four distinguished neighborhoods in $\mathcal{L}(\sigma_r)$ (see Figure 6.4). Let

$$\mathcal{U}_r = B(\gamma_1, t_1, \rho_1), \qquad \mathcal{U'}_r = B(\gamma_2, t_2, \rho_2),$$
$$\mathcal{V}_r = B(\gamma_3, t_3, \rho_3), \qquad \mathcal{V'}_r = B(\gamma_4, t_4, \rho_4),$$

where for $i \in \{1, \dots, 4\}$, γ_i is a representative of a linear approach that is represented by some x_i with $0 < x_1 < x_2 < \frac{\pi}{2} < x_3 < x_4 < \pi$ and $t_i > 0$, $\rho_i > 0$ are small enough so that the $B(\gamma_i(t_i), \rho_i)$ are disjoint from all slits v_n, from the sides of the boxes, and from each other.

Now let $n \geq 2$ be large enough and let y_n be the number which corresponds to the horizontal linear approach with base point A_n and going to the right. For $x \in \mathbb{R}_+^r$ with $x > y_n$ but sufficiently close to y_n, x is in neither of $\mathcal{U}_r$, $\mathcal{U'}_r$, $\mathcal{V}_r$, or $\mathcal{V'}_r$. When x is increased and gets closer to $y_n + x_1$ we eventually have $x \in \mathcal{U}_r$. When x increases further, it leaves $\mathcal{U}_r$ to enter $\mathcal{U'}_r$. Now there are two cases to consider.

Case 1: If $r_n = 1$, when increasing x further the linear approach will eventually be contained in the cylinder attached to v_n (with base point the lower right edge of the glued-in rectangle), and after additional time π will not be contained in the cylinder any more but have base point the tip of the slit v_n. After running for about another $\frac{\pi}{2} + x_1$, x will reenter $\mathcal{U}_r$. Then it will enter $\mathcal{U'}_r$, $\mathcal{V}_r$, $\mathcal{V'}_r$, to finally enter the same cylinder at A_n from the left side. After running for another π, x will exit the cylinder, enter $\mathcal{V}_r$ and $\mathcal{V'}_r$ again. When x reaches $y_n + 5\pi$, the linear approach will

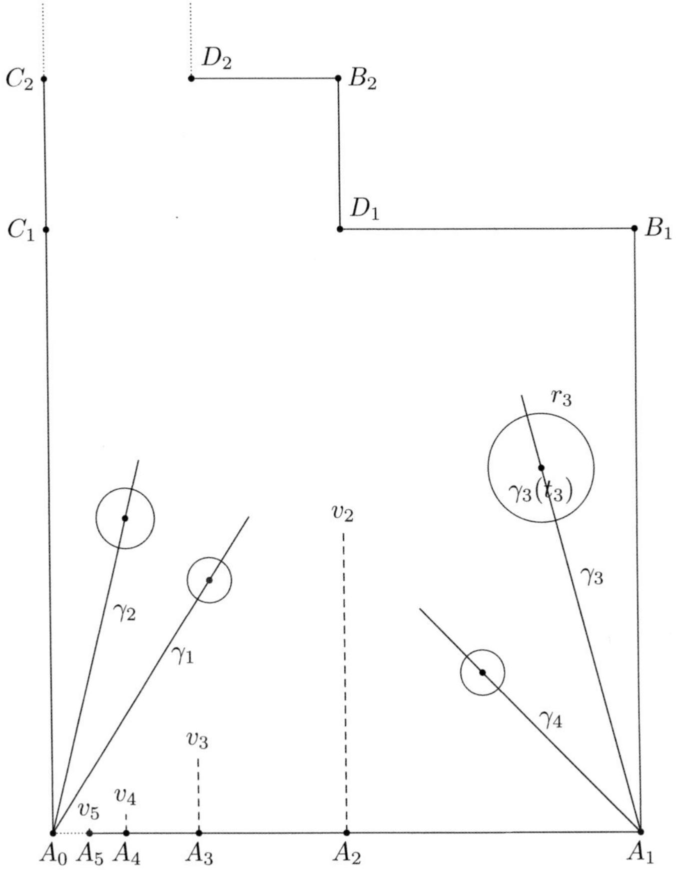

Figure 6.4: Stack of boxes with the additional vertical slits v_n (not drawn to scale in vertical direction).

leave the 0th box and enter the $(n-1)$th box from the upper right corner at B_n.

Case 2: If $r_n = 0$, after going through $\mathcal{U}_r$ and $\mathcal{U'}_r$, x will simply go through $\mathcal{V}_r$ and $\mathcal{V'}_r$ and then has base point B_n when reaching $y_n + \pi$.

By recording the successive passages in each of the four neighborhoods, writing 1 when seeing $(\mathcal{U}_r, \mathcal{U'}_r, \mathcal{U}_r, \mathcal{U'}_r, \mathcal{V}_r, \mathcal{V'}_r, \mathcal{V}_r, \mathcal{V'}_r)$ and 0 when seeing $(\mathcal{U}_r, \mathcal{U'}_r, \mathcal{V}_r, \mathcal{V'}_r)$, we record a sequence in $\mathbb{K}$ that eventually agrees with r up to shift. Note that the exact values of $0 < x_1 < x_2 < \frac{\pi}{2} < x_3 < x_4 < \pi$ (and also of the $t_1, \ldots, t_4$ and $\rho_1, \ldots, \rho_4$) do not matter for this argument. We could even choose neighborhoods of $[\gamma_1], \ldots, [\gamma_4]$ that are not elements of the subbasis defined in Lemma 1.22. In the latter case, we have to be careful to record the neighborhoods only up to multiple leavings and reentries in one neighborhood before we enter the next neighborhood.

To finish the proof, let $r, r' \in \mathbb{K}$ but not in the equivalence class of $(0, 0, 0, \ldots)$ and φ a homeomorphism from $\mathcal{L}(\sigma_r)$ to $\mathcal{L}(\sigma_{r'})$. Then φ induces a homeomorphism $f\colon \mathbb{R}^r_+ \to \mathbb{R}^{r'}_+$. We want to show that the tails of r and r' agree up to shift by looking at the recorded sequences. Hence we have to examine the images of the sets $\mathcal{U}_r, \mathcal{U'}_r, \mathcal{V}_r, \mathcal{V'}_r$ under φ which we will do by examining the images of x_1, x_2, x_3, x_4 under f.

With the arguments of the proof of Proposition 6.2 it can be shown that all elements of $(\pi, \infty) \subseteq \mathbb{R}^r_+$ have a path-connected neighborhood. On the other hand, a close look reveals that the elements in $(0, \pi] \subseteq \mathbb{R}^r_+$ do not have a path-connected neighborhood (cf. the subbasis discussion in the original version of the stack of boxes in Section 2.3). As a homeomorphism must preserve the set of points that have path-connected neighborhoods, we have $f((0, \pi]) = (0, \pi]$.

Furthermore, the neighborhoods of $\frac{\pi}{2}$ in $\mathbb{R}^r_+$ and in $\mathbb{R}^{r'}_+$ can be distinguished from the neighborhoods of other points in $(0, \pi]$. This is because the linear approach corresponding to $\frac{\pi}{2}$ has the same direction as the slits and the length of the linear approaches defined by the slits tends to 0 so the linear approaches corresponding to the slits can not be contained in a small neighborhood of the linear approach corresponding to $\frac{\pi}{2}$. Hence we have $f(\frac{\pi}{2}) = \frac{\pi}{2}$.

We show next that the map $f\colon \mathbb{R}^r_+ \to \mathbb{R}^{r'}_+$ is monotone. Let I_r be any open interval in $(0, \infty)$ of length smaller than π and equip it with the subspace topology from $\mathbb{R}^r_+$. Then this subspace topology is the same as the Euclidean topology of I_r as for every given direction there exists at most one linear approach in the corresponding subset of $\mathcal{L}(\sigma_r)$ (cf. again the subbasis discussion in Section 2.3). We choose I_r small enough so that $f(I_r) \subseteq \mathbb{R}_+^{r'}$ is contained in an open interval of length smaller than π. Then the sets I_r and $f(I_r)$ carry the Euclidean topology and the map $f_{|I_r} \colon I_r \to \mathbb{R}_+^{r'}$ is homeomorphic to its image, so it is monotone. This implies that $f \colon \mathbb{R}_+^r \to \mathbb{R}_+^{r'}$ is locally monotone and so it is monotone.

As a consequence of the monotony, we can choose x_1, x_2, x_3, x_4 with

$$\begin{array}{ccccccccccccl} 0 & < & x_1 & < & x_2 & < & \frac{\pi}{2} & < & x_3 & < & x_4 & < & \pi \quad \text{and} \\ 0 & < & f(x_1) & < & f(x_2) & < & \frac{\pi}{2} & < & f(x_3) & < & f(x_4) & < & \pi. \end{array}$$

When choosing the numbers $\rho_1, \rho_2, \rho_3, \rho_4$ sufficiently small, the neighborhoods $\varphi(\mathcal{U}_r), \varphi(\mathcal{U}'_r), \varphi(\mathcal{V}_r), \varphi(\mathcal{V}'_r)$ can serve to record the sequence in $\mathbb{K}$ and will record the same sequence as $\mathcal{U}_r, \mathcal{U}'_r, \mathcal{V}_r, \mathcal{V}'_r$. This shows that the tails of the recorded sequences in $\mathbb{K}$ for x in $\mathcal{L}(\sigma_r)$ and for $f(x)$ in $\mathcal{L}(\sigma_{r'})$ agree up to shift. Hence the tails r and r' agree up to shift and so they are in the same equivalence class, as was to be shown. □

For the translation surfaces constructed in the proof of Theorem 5, it holds that if r and r' are not in the same equivalence class then $(X_r, \mathcal{A}_r)$ and $(X_{r'}, \mathcal{A}_{r'})$ are not in the same $\mathrm{GL}^+(2,\mathbb{R})$-orbit, i.e. there exists no affine map from $(X_r, \mathcal{A}_r)$ to $(X_{r'}, \mathcal{A}_{r'})$. This is true because an affine map from $(X_r, \mathcal{A}_r)$ to $(X_{r'}, \mathcal{A}_{r'})$ would induce a homeomorphism from $\mathcal{L}(\sigma_r)$ to $\mathcal{L}(\sigma_{r'})$ (cf. Proposition 1.35).

Thus we have implicitly shown the following corollary.

Corollary 6.4 (Existence of uncountably many $\mathrm{GL}^+(2,\mathbb{R})$*-orbits).* There exist uncountably many translation structures on a Loch Ness monster of finite area which are not contained in the same $\mathrm{GL}^+(2,\mathbb{R})$-orbit.

6.3 Every finite space occurs as space of rotational components

This section is devoted to the proof of Theorem 6 asserting that any finite topological space can be realized as the space of rotational components of a translation surface. In particular, this implies that $\widetilde{\mathcal{L}}(\sigma)$ can be non-T_0, T_0 but not T_1, T_1 but not T_2, or T_2.

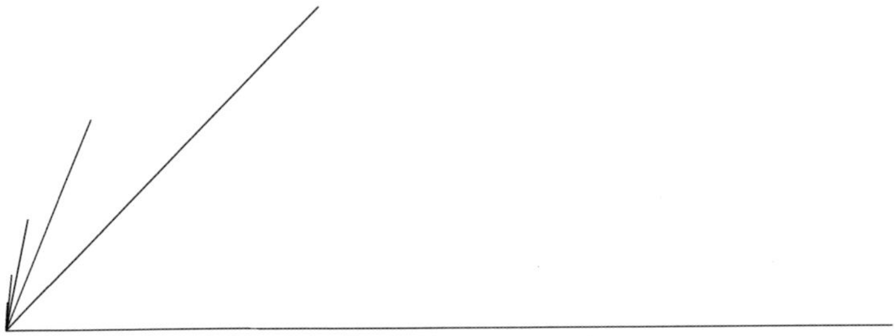

Figure 6.5: For the shrinking star decoration, we glue in rectangles at every slit.

So far, we have already seen in Chapter 2 that the space of rotational components can carry different types of topologies. For instance, if the cardinality of $\widetilde{\mathcal{L}}(\sigma)$ is 2 then the topology can be the trivial topology as for a translation surface with the geometric series decoration from Section 2.4, it can be the discrete topology as for the exponential surface from Section 2.2, or it can be the third possible topology as for a translation surface with the modified geometric series decoration from Section 2.4. However, the space of rotational components does not have to be finite. For example, the singularity of the star decoration with cylinders in Section 2.4 has a space of rotational components which is homeomorphic to S^1. When $\widetilde{\mathcal{L}}(\sigma)$ is infinite, it also can be either a Hausdorff space (cf. again the star decoration with cylinders in Section 2.4) or not (cf. the Chamanara surface in Section 2.1).

The following modification of the star decoration with cylinders from Section 2.4 will serve as a building block in the proof of Theorem 6.

Example 6.5 (Shrinking star decoration). Consider a Euclidean plane $\mathbb{R}^2$ with geodesic segments l_n from $(0,0)$ to $(2^{-n}\sin(\frac{\pi}{2^n}), 2^{-n}\cos(\frac{\pi}{2^n}))$ for $n \geq 1$ as in Figure 6.5. For every $n \geq 1$, cut open along the segment l_n and glue in a $(2+2^{-n})$-by-2^{-n} rectangle as for the star decoration in Section 2.4: glue the two smaller edges of the rectangle to the slit l_n and glue the top and bottom edges of the squares in a crosswise way as in Figure 6.6.

In this case, we have one wild singularity with one rotational component, whose translation structure is isomorphic to $[0, \infty)$.

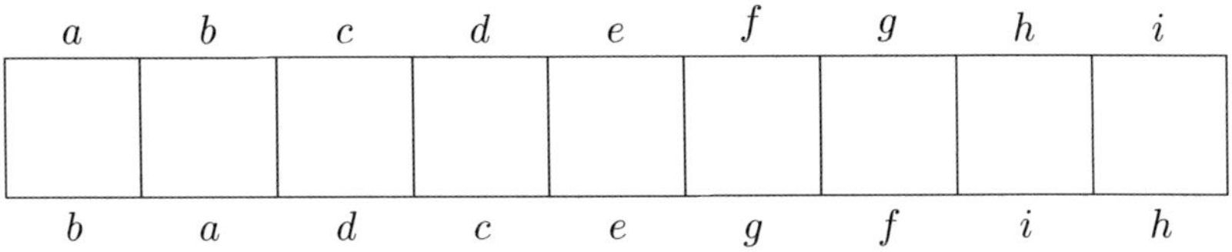

Figure 6.6: A rectangle glued in at the slit l_2: segments with the same letters are glued.

We will now use the shrinking star decoration to construct a translation surface with a given space of rotational components.

Theorem 6 (Every finite space occurs as $\widetilde{\mathcal{L}}(\sigma)$)
Let Y be a non-empty topological space of finite cardinality. Then there exists a translation surface $(X, \mathcal{A})$ with a singularity σ so that $\widetilde{\mathcal{L}}(\sigma)$ carries the same topology as Y.

Proof. Topologies on a finite set are in one-to-one correspondence with preorders "$\leq$" defined by $x \leq y$ if and only if x is in the closure of $\{y\}$. So, let $n \geq 1$ and $Y = \{y_1, \ldots, y_n\}$ be a topological space of cardinality n and "$\leq$" the corresponding preorder.

We will construct the translation surface $(X, \mathcal{A})$ explicitly starting with a Euclidean plane with n shrinking star decorations. Then we glue in additional tori along slits to identify the singularities of all different shrinking star decorations to the desired singularity σ. We obtain n rotational components. In a last step, we introduce more gluings for the rotational components that shall be comparable under the preorder to obtain that $\widetilde{\mathcal{L}}(\sigma)$ is homeomorphic to Y.

For the beginning, let us state the setting:

- Consider a Euclidean plane with n copies of the shrinking star decoration as in Example 6.5, labeled with $1, 2, \ldots, n$. The ith shrinking star decoration corresponds to $y_i \in Y$.

- For now, the shrinking star decoration labeled i has one wild singularity σ_i with one rotational component c_i which is isometric to $[0, \infty)$. The linear approach that is corresponding to 0 is called $[\gamma_i]$ with representative γ_i.

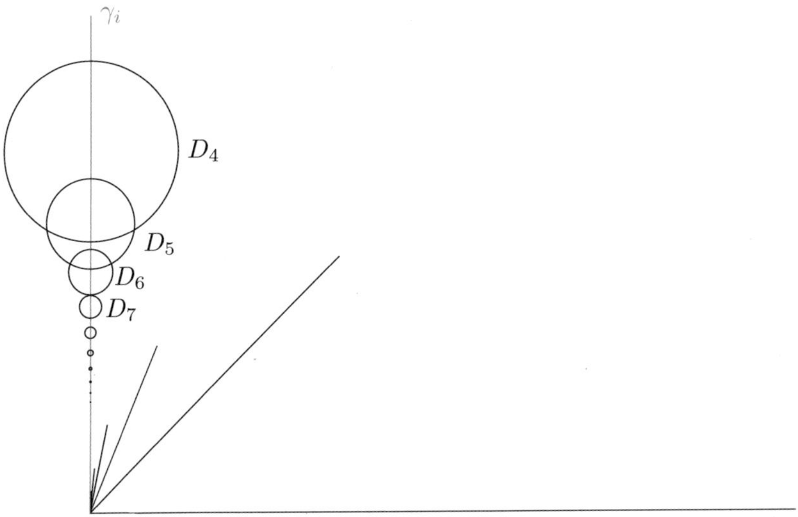

Figure 6.7: Shrinking star decoration with additional disks D_k.

- For the ith copy of the shrinking star decoration and every $k \geq 1$, let D_k be the disk with center $\gamma_i(\frac{1}{k})$ and radius 2^{-k} as shown in Figure 6.7. This gives us an infinite sequence of disks where all except the first six disks are disjoint. We will do all gluings only for $k \geq 6$ without explicitly mentioning this restriction every time.
- Recall that for the kth branch of each copy of the shrinking star decoration, there is a rectangle as in Figure 6.6 glued to the two sides of the branch. By the *slit in the kth rectangle*, we mean the geodesic segment which corresponds to the glued top and bottom of the middle square of the rectangle (labeled by e in Figure 6.6). We will cut open and reglue along some of these slits later.
- Let T_k be a torus, glued from a square of side length 2^{-k} for each $k \geq 1$.
- For $i, j \in \{1, \ldots, n\}$ define $a_k^{i,j} = kn^2 + in + j$ for every $k \geq 1$. This gives n^2 infinite sequences in $\mathbb{N} \setminus \{0\}$ so that no two sequences have a common element.

Now we glue the copies of the shrinking star decoration in such a way that all singularities σ_i are identified to one singularity σ. For every $k \geq 1$ and every $i \in \{1, \dots, n\}$, glue the slit in the $a_k^{i,i}$th rectangle of the ith shrinking star decoration to a suitable slit, i.e. a slit with the same holonomy vector, on the kth torus. As the side length 2^{-k} of the tori is shrinking, the slits come the closer to each other the greater k is. This makes sure that $d(\sigma_i, \sigma_j) = 0$ and hence $\sigma_i = \sigma_j$ for all $i, j \in \{1, \dots, n\}$. The resulting singularity is called σ. Furthermore, the additional gluings extend the length of the rotational components but the topology of $\widetilde{\mathcal{L}}(\sigma)$ is the discrete topology for now (cf. the arguments in the proof of Proposition 6.2).

We have to do more gluings to provide $\widetilde{\mathcal{L}}(\sigma)$ with the topology of Y. If $i \neq j$ and $y_i \leq y_j$, i.e. if every open set in Y containing y_i contains y_j, then we glue the slit in the $a_k^{i,j}$th rectangle of the jth shrinking star decoration to a suitable slit inside the interior of the right half of the $a_k^{i,j}$th disk on the ith shrinking star decoration for every $k \geq 1$. These gluings extend the length of the rotational component c_j but do not change the length of the rotational component c_i. More important, for every $t > 0$ and $r > 0$ there exists a linear approach contained in $B(\gamma_i, t, r)$ which starts at the end point of the slit in the $a_k^{i,j}$th disk for some $k \geq 1$. This linear approach is contained in the rotational component c_j and hence every open neighborhood of $[\gamma_i]$ has nonempty intersection with the set of linear approaches contained in c_j. This means that every open neighborhood of c_i in $\widetilde{\mathcal{L}}(\sigma)$ contains c_j.

This ends the construction and all in all, we have shown that $\widetilde{\mathcal{L}}(\sigma)$ is homeomorphic to Y as was to be proven. □

Bibliography

[AHar] Artur Avila and Pascal Hubert, *Recurrence for the wind-tree model*, Annales de l'Institut Henri Poincaré (C) Analyse Non Linéaire (to appear).

[BH99] Martin R. Bridson and André Haefliger, *Metric spaces of non-positive curvature*, Die Grundlehren der mathematischen Wissenschaften in Einzeldarstellungen; 319, Springer, Berlin, 1999.

[Bow12] Joshua P. Bowman, *Finiteness conditions on translation surfaces*, Quasiconformal mappings, Riemann surfaces, and Teichmüller spaces. AMS special session in honor of Clifford J. Earle, Syracuse, NY, USA, October 2–3, 2010, American Mathematical Society, Providence, RI, 2012, pp. 31–40.

[Bow13] ______, *The complete family of Arnoux-Yoccoz surfaces*, Geometriae Dedicata **164** (2013), 113–130.

[BV13] Joshua P. Bowman and Ferrán Valdez, *Wild singularities of flat surfaces*, Israel Journal of Mathematics **197** (2013), 69–97.

[Cha04] Reza Chamanara, *Affine automorphism groups of surfaces of infinite type*, In the Tradition of Ahlfors and Bers, III (William Abikoff and Andrew Haas, eds.), Contemporary mathematics, vol. 355, 2004, pp. 123–145.

[CRW14] Lucien Clavier, Anja Randecker, and Chenxi Wu, *Rotational component spaces for infinite-type translation surfaces*, arxiv:1412.0633, 2014.

[DH07] Max Dehn and Poul Heegaard, *Analysis situs*, Enzyklopädie der mathematischen Wissenschaften III_I, 153-220, 1907.

[DHL14] Vincent Delecroix, Pascal Hubert, and Samuel Lelièvre, *Diffusion for the periodic wind-tree model*, Annales Scientifiques de l'École Normale Supérieure. Quatrième Série **47** (2014), no. 6, 1085–1110.

[Dyc88] Walther Dyck, *Beiträge zur Analysis situs. I*, Mathematische Annalen **32** (1888), 457–513.

[EM13] Alex Eskin and Maryam Mirzakhani, *Invariant and stationary measures for the* $\mathrm{SL}(2,\mathbb{R})$ *action on moduli space*, arXiv:1302.3320, 2013.

[EMM15] Alex Eskin, Maryam Mirzakhani, and Amir Mohammadi, *Isolation, equidistribution, and orbit closures for the* $\mathrm{SL}(2,\mathbb{R})$ *action on moduli space*, Annals of Mathematics. Second Series **182** (2015), no. 2, 673–721.

[FK36] Ralph H. Fox and Richard B. Kershner, *Concerning the transitive properties of geodesics on a rational polyhedron*, Duke Mathematical Journal **2** (1936), 147–150.

[FM13] Giovanni Forni and Carlos Matheus, *Introduction to Teichmüller theory and its applications to dynamics of interval exchange transformations, flows on surfaces and billiards*, arXiv:1311.2758, 2013.

[Fre31] Hans Freudenthal, *Über die Enden topologischer Räume und Gruppen*, Mathematische Zeitschrift **33** (1931), no. 1, 692–713.

[FU14] Krzysztof Frączek and Corinna Ulcigrai, *Non-ergodic* $\mathbb{Z}$*-periodic billiards and infinite translation surfaces*, Inventiones mathematicae **197** (2014), no. 2, 241–298.

[Ghy95] Etienne Ghys, *Topologie des feuilles génériques*, Annals of Mathematics. Second Series **141** (1995), no. 2, 387–422.

[GX13] Jean Gallier and Dianna Xu, *A guide to the classification theorem for compact surfaces*, Geometry and Computing, vol. 9, Springer, Berlin, 2013.

[HHW13] W. Patrick Hooper, Pascal Hubert, and Barak Weiss, *Dynamics on the infinite staircase*, Discrete and Continuous Dynamical Systems **33** (2013), no. 9, 4341–4347.

[HLT11] Pascal Hubert, Samuel Lelièvre, and Serge Troubetzkoy, *The Ehrenfest wind-tree model: periodic directions, recurrence, diffusion*, Journal für die Reine und Angewandte Mathematik **656** (2011), 223–244.

[Hoo13a] W. Patrick Hooper, *Immersions and the space of all translation structures*, arXiv:1310.5193, 2013.

[Hoo13b] ———, *Immersions and translation structures on the disk*, arXiv:1309.4795, 2013.

[Hoo14] ———, *An infinite surface with the lattice property I: Veech groups and coding geodesics*, Transactions of the American Mathematical Society **366** (2014), no. 5, 2625–2649.

[Hoo15] ———, *The invariant measures of some infinite interval exchange maps*, Geometry & Topology **19** (2015), no. 4, 1895–2038.

[Hop43] Heinz Hopf, *Enden offener Räume und unendliche diskontinuierliche Gruppen*, Commentarii Mathematici Helvetici **16** (1943), 81–100.

[HS06] Pascal Hubert and Thomas A. Schmidt, *An introduction to Veech surfaces*, Handbook of dynamical systems. Volume 1B, Elsevier, Amsterdam, 2006, pp. 501–526.

[HS10] Pascal Hubert and Gabriela Schmithüsen, *Infinite translation surfaces with infinitely generated Veech groups*, Journal of Modern Dynamics **4** (2010), no. 4, 715–732.

[HW12] W. Patrick Hooper and Barak Weiss, *Generalized staircases: recurrence and symmetry*, Annales de l'Institut Fourier **62** (2012), no. 4, 1581–1600.

[HW13] Pascal Hubert and Barak Weiss, *Ergodicity for infinite periodic translation surfaces*, Compositio Mathematica **149** (2013), no. 8, 1364–1380.

[Jor66] Camille Jordan, *La deformation des surfaces*, Journal de Mathématiques Pures et Appliquées. Deuxième série **11** (1866), 105–109.

[Ker23] Béla v. Kerékjártó, *Vorlesungen über Topologie. I. Flächentopologie*, Die Grundlehren der mathematischen Wissenschaften in Einzeldarstellungen; Bd. 8, Springer, Berlin, 1923.

[KMS86] Steven Kerckhoff, Howard Masur, and John Smillie, *Ergodicity of billiard flows and quadratic differentials*, Annals of Mathematics. Second Series **124** (1986), 293–311.

[Mas86] Howard Masur, *Closed trajectories for quadratic differentials with an application to billiards*, Duke Mathematical Journal **53** (1986), 307–314.

[Mas06] ______, *Ergodic theory of translation surfaces*, Handbook of dynamical systems. Volume 1B, Elsevier, Amsterdam, 2006, pp. 527–547.

[MT02] Howard Masur and Serge Tabachnikov, *Rational billiards and flat structures*, Handbook of dynamical systems. Volume 1A, North-Holland, Amsterdam, 2002, pp. 1015–1089.

[Mö63] August Ferdinand Möbius, *Theorie der elementaren Verwandtschaft*, Berichte über die Verhandlungen der königlich sächsischen Gesellschaft der Wissenschaften zu Leipzig, mathematisch-physische Classe **15** (1863), 18–57.

[Poi90] Henri Poincaré, *Sur le problème des trois corps et les équations de la dynamique*, Acta Mathematica **13** (1890), 1–270.

[PS81] Anthony Phillips and Dennis Sullivan, *Geometry of leaves*, Topology **20** (1981), 209–218.

[PSV11] Piotr Przytycki, Gabriela Schmithüsen, and Ferrán Valdez, *Veech groups of Loch Ness monsters*, Annales de l'Institut Fourier **61** (2011), no. 2, 673–687.

[Rad25] Tibor Radó, *Über den Begriff der Riemannschen Fläche*, Acta Litterarum ac Scientiarum Szeged **2** (1925), 101–121.

[Ran12] Anja Randecker, *Skript zur Vortragsreihe „Unendliche Translationsflächen“*, Lecture notes, 2012.

[Ran14] ———, *When wild singularities of translation surfaces imply infinite genus*, arXiv:1410.1501, 2014.

[Rat94] John G. Ratcliffe, *Foundations of hyperbolic manifolds*, Graduate texts in mathematics; 149, Springer, New York, 1994.

[Ric63] Ian Richards, *On the classification of noncompact surfaces*, Transactions of the American Mathematical Society **106** (1963), no. 2, 259–269.

[Sch04] Gabriela Schmithüsen, *An algorithm for finding the Veech group of an origami*, Experimental Mathematics **13** (2004), no. 4, 459–472.

[Str84] Kurt Strebel, *Quadratic differentials*, Ergebnisse der Mathematik und ihrer Grenzgebiete; Folge 3, Bd. 5, Springer, Berlin, 1984.

[Thu81] William P. Thurston, *The geometry and topology of three-manifolds*, Lecture Notes, Princeton University, 1978–1981.

[Thu88] ———, *On the geometry and dynamics of diffeomorphisms of surfaces*, Bulletin of the American Mathematical Society. New Series **19** (1988), no. 2, 417–431.

[Tre14] Rodrigo Treviño, *On the ergodicity of flat surfaces of finite area*, Geometric and Functional Analysis **24** (2014), no. 1, 360–386.

[Val09] Ferrán Valdez, *Infinite genus surfaces and irrational polygonal billiards*, Geometriae Dedicata **143** (2009), 143–154.

[Val12] ———, *Veech groups, irrational billiards and stable Abelian differentials*, Discrete and Continuous Dynamical Systems **32** (2012), no. 3, 1055–1063.

[Vee89] William A. Veech, *Teichmüller curves in moduli space, Eisenstein series and an application to triangular billiards*, Inventiones Mathematicae **97** (1989), no. 3, 553–583.

[Vee90] William A. Veech, *Moduli spaces of quadratic differentials*, Journal d'Analyse Mathématique **55** (1990), 117–171.

[Vor96] Yaroslav Vorobets, *Planar structures and billiards in rational polygons: the Veech alternative*, Russian Mathematical Surveys **51** (1996), no. 5, 779–817.

[Zie08] Günter M. Ziegler, *Polyhedral surfaces of high genus*, Discrete differential geometry, Birkhäuser, Basel, 2008, pp. 191–213.

[ZK76] Aleksandr N. Zemlyakov and Anatole B. Katok, *Topological transitivity of billiards in polygons*, Mathematical notes of the Academy of Sciences of the USSR **18** (1976), no. 2, 760–764.

[Zor06] Anton Zorich, *Flat surfaces*, Frontiers in number theory, physics, and geometry I. On random matrices, zeta functions, and dynamical systems. Papers from the meeting, Les Houches, France, March 9–21, 2003, Springer, Berlin, 2nd printing ed., 2006, pp. 437–583.

Index

T

U

V

X